Praise for

inconspicuous co

T0013774

"Focusing on food, fashion, technology, and fuel, she shows how even the smallest decisions can have profound environmental consequences." —*New York Times*

"A compelling—and illuminating—look at how our daily habits impact the environment…[Schlossberg's] wry, sometimes self-deprecating humor makes the depth of research and information provided throughout the book go down easy." —*Vanity Fair*

"*Inconspicuous Consumption* is scary informative—in both senses—but also oddly enjoyable, filled with salty jokes and fun (or not so fun) facts.…If you're looking for something to cling to in what often feels like a hopeless conversation, Schlossberg's darkly humorous, knowledge-is-power, eyes-wide-open approach may be just the thing." —*Vogue*

"Entertaining and eye-opening…the sharp, well-written book doesn't read like an admonishment; instead, it's a call to action that reminds us all of our responsibility and capability to change the world." —*Town & Country*

"The subject of climate change is inescapable, as it should be, but too few stories focus on one's everyday impact upon the environment. In *Inconspicuous Consumption*, former *New York Times* science writer Tatiana Schlossberg breaks down exactly how everyday activities—watching Netflix, eating a burger, turning on the light—impact the environment." —*Bustle*

"Schlossberg adeptly guides readers toward understanding the unlikely implications of how the manufacture of everyday acquisitions...exact environmental and human costs. Beyond individual choices, though, Schlossberg's sophisticated understanding of the world's complexity and her conversational style rally readers to vigilance about corporate and governmental oversight in this small world." —*National Book Review*

"The author breaks complex issues down to be understandable to the lay reader, while her humor and wit ensure that readers will close the book feeling energized rather than hopeless." —*Booklist* (starred review)

"Schlossberg brings a variety of current conversations on environment together in down-to-earth, easily understood terms. Avoiding dense technical language and writing in a highly personalized style laced with humor and asides, the author provides much-needed clarifications about climate change and pollution that not only empower average consumers with the ability to act and make informed decisions, but also encourage and inspire that action. If fighting climate change can be engaging, fun, and fulfilling, this is the road map." —*Kirkus Reviews*

"[A] straightforward, accessible look at the environmental impact of consumer habits...With insight and urgency, Schlossberg prods readers to think more deeply...[and] delivers an intriguing and educational narrative." —*Publishers Weekly*

"With this call for mass action, [Schlossberg] presents valuable information that could help readers make more sustainable choices in their lives." —*Library Journal*

inconspicuous consumption

the environmental impact you don't know you have

tatiana schlossberg

GRAND CENTRAL
PUBLISHING

NEW YORK   BOSTON

Grand Central Publishing
Hachette Book Group
1290 Avenue of the Americas, New York, NY 10104
grandcentralpublishing.com
twitter.com/grandcentralpub

Originally published in hardcover and ebook by Grand Central Publishing in August 2019
First trade paperback edition: March 2022

Grand Central Publishing is a division of Hachette Book Group, Inc.
The Grand Central Publishing name and logo is a trademark of
Hachette Book Group, Inc.

The publisher is not responsible for websites (or their content)
that are not owned by the publisher.

The Hachette Speakers Bureau provides a wide range of authors for speaking events.
To find out more, go to www.hachettespeakersbureau.com or call (866) 376-6591.

Library of Congress Cataloging-in-Publication Data

Names: Schlossberg, Tatiana, author.
Title: Inconspicuous consumption : the environmental impact you don't know you have /
Tatiana Schlossberg.
Description: First edition. | New York : Grand Central Publishing, 2019. |
Includes bibliographical references.
Identifiers: LCCN 2019007474 | ISBN 9781538747087 (hardcover) | ISBN
9781538747094 (ebook)
Subjects: LCSH: Consumption (Economics)—Environmental aspects. | Consumer
behavior—Environmental aspects. | Lifestyles—Environmental aspects. |
Environmental responsibility.
Classification: LCC HC79.C6 S258 2019 | DDC 363.7—dc23
LC record available at https://lccn.loc.gov/2019007474

ISBNs: 978-1-5387-4707-0 (trade paperback), 978-1-5387-4709-4 (ebook)

Printed in the United States of America

LSC-C

Printing 1, 2022

To all of the climate scientists, environmental lawyers, activists and advocates, for their help and for not giving up
And to my family

Contents

Preface

In the months after this book was originally published, I traveled around the country to talk about my book: Why I thought my book was different, what I had learned while writing it, and—most importantly to people who came to see me speak—what they could do about climate change.

Everywhere I went, I told people that the most important thing they could do if they cared about climate change and wanted to make a difference is to vote and to get involved in the political and civic processes. Over and over again. (This was also in late 2019, so the stakes for the next election could not have been higher.) I always talked about collective action, as opposed to individual behavioral changes, which I sensed people were expecting and wanted to hear about. I heard myself repeating: we should not feel individually guilty about climate change; we should feel collectively responsible for building a better world. I continue to believe this is true, no matter how many times I say it. I think people listened, but then they often still wanted to know if they should stop eating red meat or drinking from plastic water bottles, or change whatever small-scale individual behavior they had heard about most recently.

As the schedule of events surrounding the book's release

began to dwindle, the entire world changed. In China and then in Italy, reports of a new disease took over the news. Within weeks, COVID-19 came to New York, where I live. I won't recount what happened next, since you likely lived through it, too. But one aspect of it that I do feel compelled to write about is the connections people started to make to climate change, almost immediately. As cities and countries went into lockdown, people stopped commuting, factories closed or reduced their output, and air pollution seemed to disappear from cities like Los Angeles and Delhi. Ship traffic and air travel decreased dramatically, now that people were staying home and that buying new things to bring home seemed like an immediately dangerous proposition. People told me that maybe, actually, the pandemic would be "good" for climate change: it would reduce greenhouse gases and endless consumption and waste.

The eternal contrarian that I am, I saw a different story. Infinite consumption has always been dangerous to the long-term health of the planet and all of the people who live here (us). And I'm not assigning some kind of visionary status to myself here, either—most climate activists and reporters who cover the subject saw the danger in this narrative, too.

There were a few reasons why I was skeptical.

Some disagree about what exactly caused the COVID-19 pandemic, but most epidemiologists think that it jumped from some kind of wild animal population to humans. More contact of this kind between humans and animals could mean more zoonotic diseases (passing from animals to humans). In fact, this is already happening and is exacerbated by climate change, but that's not the only cause. One consequence of population growth and increasing urbanization around the world is habitat fragmentation, which describes the process by which animal habitats—where they live, but also the corridors they depend upon for migration—are broken up by human settlement and infrastructure, including buildings, roads, utilities, and anything manmade that takes up a lot of space, or something that changes the way an ecosystem functions, like draining a wetland for construction. As human settlement expands, it increasingly comes into contact with animals and plants. For example, as people move closer to grassy or forested areas, they also move into the favored habitat of ticks. Or, if people move into a new area and don't have enough food or are newly cut off from what they normally eat, they may also hunt wild animals that could carry some new diseases we haven't been exposed to yet. As temperatures warm, animals like insects and arachnids (e.g., mosquitoes and ticks) can expand their ranges, and these animals often carry diseases that they can easily pass to humans. It's not the same as a pandemic, but diseases like Lyme, malaria, dengue, and more will likely affect more people as a result of climate change.

As for the air pollution stuff, I was skeptical of that, too. It seemed like whatever emissions reductions we saw wouldn't last because these reductions weren't achieved in a sustainable way. I expected production and travel would ramp up and probably over-compensate for whatever savings we had gained as soon as it was safe to do so. And they did, because those reductions required a near-total shutdown of the global economy, which isn't possible to maintain long-term for a meaningful shift in greenhouse gas emissions. And even if it had lasted, it was, of course, not worth the pain: the death, illness, grief, and sorrow, as well as the financial hardship caused to millions of people who lost their jobs, and the children all over the world who missed out on learning and socialization and just getting to play and be kids. We don't know what the long-term effects of all of that will be.

At the peak of global lockdown regimes, greenhouse gas emissions fell by about 17 percent, according to a study published in

Nature, but were expected to end up being only about 5 percent less than they otherwise would have been over the course of the year. According to the National Oceanic and Atmospheric Administration, carbon dioxide levels in the atmosphere rose in 2020 by slightly less than in recent years—by about 1.8 parts per million, as opposed to around 2 parts per million previously (and that difference falls within the range of natural variability). We breached the global threshold of 400 parts per million in 2015, and by 2019 had crossed over to 410 parts per million. "Such a rate of increase has never been seen in the history of our records," Petteri Taalas, the secretary general of the World Meteorological Association told the BBC. In fact, the average surface temperature of the earth may have actually warmed briefly during lockdown because particulate pollution (which decreased with a lack of travel and factory output) normally reflects some heat back to space. With less particulate matter in the atmosphere, warming may have temporarily increased by 0.03°C. This impact was short-lived and tiny, but it was actually bigger than any emissions reductions by lockdown restrictions.

That may all sound discouraging—that many of us stayed home and stopped commuting and traveling and went through all of this, and still emissions continued on a dangerous course. And if even with a shutdown of the global economy, we could only cut emissions by 5 percent, what hope do we have?

That, of course, would be the wrong message to take away. We learned that we *can* actually change our behavior pretty quickly in response to a perceived external threat. We learned that people need support to be able to make that kind of change, but it can be done. A corollary to that: in the United States, the people who needed the most help and support largely did not get it, or did not get enough of it. If we want people to be able to make the changes required—for a pandemic or climate change—we have to support, as a society, those who need help; we won't survive as a society without doing so.

Social distancing was logistically possible for people who could work from home. For "essential workers"—those whom society depends on for everything from getting food to picking up trash to cleaning hospitals (not to mention doctors, nurses, and EMTs) but who are rarely treated or paid as if they are "essential" to everyone else—that was not the case. These were mainly working-class people, primarily from low-income communities and communities of color. The disproportionate exposure and suffering experienced by these communities are consequences of structural racism and inequality. These communities suffered from COVID-19 at higher rates and were more likely to be hospitalized and to die.

Unsurprisingly, there is a connection to climate change and the environment there, too.

One of the most harmful pollutants that results from burning fossil fuels is small particulate matter, or PM 2.5. PM 2.5 is like soot, but it's small enough that it can get into your lungs and your bloodstream. Long-term exposure to PM 2.5 can cause asthma, lung disease, heart disease, low birth weight, and other developmental problems. It's responsible for 85,000 to 200,000 excess deaths in the United States every year. One study found that PM 2.5 accounted for around 8.7 million premature deaths in 2018, with nearly 5 million in China and India combined.[1] A study published in April 2021 showed that race is the most significant determinant of exposure to PM 2.5 in the United States: compared to white Americans, communities of color are more exposed to PM 2.5 from all sources, with Black Americans most affected. Long-term exposure to PM 2.5 is also correlated with hospitalization for COVID-19.[2] (Recently, we also learned that the massive western wildfires in 2020, which released significant amounts of PM 2.5

and other pollutants, may have contributed to thousands of additional COVID-19 cases and deaths. The wildfires we are seeing in the American West are made more frequent and intense because of the warming temperatures caused by climate change.[3])

The injustices are overwhelming and infuriating. But they shouldn't be surprising: climate change is fundamentally a justice issue.

In the last year, the world also convulsed with upheavals related to police brutality and racial justice. These protests came, in part, in response to the murder of George Floyd by Derek Chauvin, a police officer, as well as to the killings of many other Black people by the police, and to the systemic inequality and racism these killings reflect. The protests that took place in cities across America and around the world built on years of work and commitment by members of the Black Lives Matter movement and other civil rights activists.

According to a survey from the Yale Program on Climate Change Communication, Black communities and Latinx communities already care about climate change and climate action more than white communities. Black people, Indigenous people, and Latinx communities, primarily, have been protesting pollution and waste in their neighborhoods and communities, as well as climate change, for decades. Members of these communities are some of the original environmental activists, even though the mainstream environmental movement, largely dominated by white activists, has not always recognized their contributions. The predominantly white environmental movement needs to participate even more in the crucial work of environmental and climate justice and follow and support these leaders. After all, they are experts on environmental degradation and climate change already, by virtue of the social, environmental, and climate injustices that have affected their communities. It is not possible to create a safe and habitable planet without creating a just and equal society.

Climate justice is a global issue, too. Around the world, activists from the Global South—particularly in parts of Africa, South Asia, and Pacific Island nations—have been calling attention to the injustices of climate change, too, and the need for Western countries to redress them. These countries (largely in the Global North) are more responsible for climate change (because of their historic greenhouse gas emissions) than those in the Global South, which will be affected first and worst.

I wrote about the issues of climate and environmental justice in this book, particularly in the United States and particularly in the chapter about fuel, but they appear throughout the book. I wish I had written about them all more. At its heart, climate change is a justice issue. It exacerbates inequality within and between countries; its effects are experienced unequally. To care about climate change is to care about justice and equality. Climate action without a relentless commitment to justice and fairness isn't real climate action. We need to save the polar bears, sure, but we also need to save the people. Actually, we can't save the polar bears if we don't save the people.

But there's much more I could have written, and there is important work on the subject from people with much more authority, knowledge, and expertise than me. I encourage you to find and follow their work, which is where I have learned and continue to learn about these issues. Kendra Pierre-Louis, Rhiana Gunn-Wright, Mary Annaïse Heglar, Julian Brave NoiseCat, Elizabeth Yeampierre, Robert Bullard, Heather McTeer Toney, Catherine Coleman Flowers, Yessenia Funes, Thimali Kodikara, and Gerald Torres are a few of the writers, advocates, and thinkers whose work on this subject (and others) I admire.

The pandemic and the fight for racial justice and equality both have made it clear that protecting our health and our ability to live together in society without fear and with confidence in our system of laws is unfinished work that requires everyone's help. You'll notice as you read this book that there aren't a lot of simple "solutions" offered, like replacing your plastic water bottle or whatever. Individual behavioral changes are great, but they won't be enough to fight climate change because, like addressing systemic injustice, preserving the habitability of our planet is the urgent work of all people on a structural level.

About a year after this book was published, the Society of Environmental Journalists decided to award it the Rachel Carson Environment Book Award. This was an honor I dreamt not of! But I am very grateful and awed to have received it. It has made me think more, too, about Rachel Carson, without whose work the environmental movement might never have taken off. Shortly before her death from cancer in 1964, she appeared on a CBS broadcast about her world-shaking book, *Silent Spring*. In that program, she said, "Man's attitude toward nature is today critically important simply because we have now acquired a fateful power to alter and destroy nature. But man is a part of nature, and his war against nature is inevitably a war against himself. [We are] challenged as mankind has never been challenged before to prove our maturity and our mastery, not of nature, but of ourselves."

It's hard to want to keep writing after I read Rachel Carson (Why bother? It's so much better than what I can come up with), but unfortunately, I haven't quite finished this foreword. I've tried to write a book that explains some parts of the climate crisis: why we're in this mess, and why it's so hard to get out of. It's important to understand a problem so that you can try to solve it, rather than letting it overwhelm you.

We also don't have a lot of time to waste. I generally resist the framing of "we only have XX years left to stay within 1.5°C of temperature increase" because I think it makes people think there just isn't any chance at all. Change the way the entire global economy works in less than a decade? Why bother?

But here's why. We know what happens if we do nothing: things get worse. But we don't know what happens if we decide to do something, especially if we decide to do a lot. The biggest variable in the trajectory of climate change is what people decide to do about it. So in that way, every little thing counts, because it's all part of doing something.

There are also millions of people already suffering from the effects of climate change. They are disproportionately Black, Brown, and Indigenous people, particularly women, particularly in the Global South. To do nothing means that we, as a global society, have decided that we're okay with that profound injustice. Even if the future weren't at stake, too, that should be enough of a reason.

And even though climate change is clearly a problem and you'll never hear me say otherwise, I hope you can also try to think about it as an opportunity to do things differently. If we live in a country and a world where we don't burn fossil fuels for energy, we will live in a world where people don't get sick and die because of that pollution. If we stop making plastic from oil and gas, we won't have to worry about what happens when plastic gets into the ocean (and everywhere else) and we can focus on getting out the plastic that's already there. We don't have to keep living like this, in fear of the future and with guilt about the past, because change is possible. It just takes work, but I personally think (and you can quote me on this) it will be worth it.

Introduction

When we think about climate change, melting polar ice caps, hurricanes, or forest fires might be the first things that come to mind. If we think a little longer, we might get all the way to renewable energy, greenhouse gas emissions, or coal. Much lower down on the list, if it comes up at all, is average, everyday, run-of-the-mill stuff, including literal stuff: a pair of jeans, a hamburger, Netflix, an air-conditioner.

But those four things, and many others, should be much higher on the list. In fact, almost everything we do, use, and eat in the United States (and much of the rest of the world) has a lot to do with climate change and the environment, because of the way we use resources, create waste, and emit greenhouse gases without even thinking about it.

That's why I wanted to write this book: the physical things we interact with every day and lots of our daily activities don't exist in a vacuum—they're much more connected to each other, to global climate change, and to each one of us than we think.

The story of climate change—and all of our stuff—is actually a story about everything: science, health, injustice, inequality, national and international politics, the natural world, business, normal life. Climate change affects everyone constantly, but, until very recently, we usually only talked about it for a few days when some natural disaster happened or a particularly scary report by government scientists came out—if then—before we moved on to something else. Really, we should be talking about it all the time. But it's scary, and even though it's "an existential crisis facing humanity," it doesn't always seem to connect to our lives, so we haven't talked about it nearly enough.

Let me back up for a second: before I became a climate change and environmental journalist, I understood in broad strokes what climate change was and why it was happening—that transportation, industry, agriculture, and electricity generation all involve processes that result in the emissions of carbon dioxide or other greenhouse gases, which trap heat in the atmosphere, causing the overall surface temperature of the planet to increase. This leads to the melting of the polar ice caps and the rise in sea level, stronger storms, drought, forest fires, flooding, etc. I knew that there were other kinds of pollution that created environmental problems, like ocean plastic and acid rain. But I didn't seek out too much more information. I didn't like reading about climate change and its effects—it filled me with dread and made me feel powerless. The problems seemed too big and too inevitable for me to do anything about, so it felt like it was probably best to look away. Intellectually, I recognized that climate change is the most important issue in the world. Pretending it wasn't happening wouldn't make it go away, so eventually I thought I should probably learn more about it, and I was lucky that the editors at the *New York Times* thought I was up to the challenge.

But it also became pretty clear to me as I started reporting and writing about climate change and the environment that there was so much left out of the conversation, including what might make the issues relatable. Plus, articles about climate change can be really boring. Even I think they're boring! They tend to be incredibly technical or presume that we all have a lot of background knowledge and context, which we don't necessarily. And that's too bad, because these issues are actually really interesting. Sure, they're complicated, but they connect to each other and to us in lots of surprising and fascinating ways.

I also noticed that it was really hard to bring climate change down to scale, to make sense of it within the context of our own lives, and to get a sense of how our habits and the products we use affect the environment. I started thinking: What kinds of things am I doing or buying without thinking about where they come from or what kind of impact they have? How have our habits and our expectations changed over time, maybe generating more waste or encouraging more consumption? What do I notice myself and other people doing that seems wasteful but appears to happen without a second thought?

I thought about how watching TV and movies is so different now from when I was little, when a show was on once a week, and if you missed it, the best you could hope for was a rerun someday. Now, I can watch a whole series in one sitting on my laptop, and streaming online videos is something that many of us probably take for granted. Maybe we think about the electricity needed to power our laptop, but we probably don't consider that going online itself uses electricity, which often comes from fossil fuels. In the US, we still get about one-third of our electricity from coal, so streaming your online video may be coal-powered; the by-product of burning that coal is coal ash, one of the largest industrial solid-waste streams in the country, which is largely under-regulated and can end up polluting groundwater, streams, lakes, and rivers across the country.[1] As crazy as this might sound, it means that watching your favorite episode of *The Office* might come at the expense of clean water for someone else.

I thought about cotton. Sometimes, we hear about how drought and irrigation for agriculture create environmental problems or freshwater shortages, but we don't always hear those same things about cotton, even though it's also a plant. Cotton has to be grown somewhere, and depending on where it comes from, as many as 2,000 gallons of water could have been used to grow one kilogram of it, and up to 2,900 gallons could have been used to make it into a pair of blue jeans, possibly affecting someone's water supply somewhere, maybe that of a villager in Uzbekistan.[2]

If you've ever looked at a guide about how to reduce your carbon footprint, you've probably read that cutting red meat out of your diet is a pretty effective way to do so.[3] But maybe there are problems that agriculture poses beyond just greenhouse gas emissions, I thought. Turns out, there are: in the US, the majority of cows mostly eat feed derived from corn and soybeans, and the way we grow those crops also causes massive amounts of water pollution—in rivers, lakes, the Gulf of Mexico, and rural drinking supplies.

For the most part, these aren't the kinds of environmental problems we hear about, but they help us understand the broader scope of the issue once we learn about them. As I began to learn more about the many aspects of climate change, I saw that these discrete threads are all actually part of a giant tapestry telling a story about pollution and waste and also about people and culture and history. It didn't make me feel less alarmed, but I did start to feel less anxious and scared. I started to feel that I understood what was happening in the world, that I could evaluate what politicians and scientists and business leaders were saying, and I found that I felt like a more informed and more responsible citizen.

Unfortunately, knowledge doesn't necessarily change our very real feelings of powerlessness. While we know what needs to happen to mitigate the effects of climate change (halting greenhouse gas emissions now and hopefully sucking some more out from the atmosphere), actually getting there is really hard, because greenhouse gas emissions are involved in almost everything we do and a certain amount of warming and change is unavoidable at this point. Few people have offered understandable solutions. We aren't told very much about either these solutions' effectiveness or the actual effort or expense involved in making them happen. Instead, we hear a lot about smaller-scale fixes that play to our individual desire to have an impact while the larger systemic problems are often left unexplained and unsolved. Most of us know that plastic bottles are wasteful and that we should drink from reusable ones instead. Or we are often told that we need to get all of our electricity from renewable sources within the next decade or so, which would require a major retooling of the economy, a transformation of the electrical grid, and developments in battery technology, but those undertakings are often treated as an afterthought. Somehow, it seems like these two options are given the same amount of consideration and attention. (Actually, the plastic bottle one probably gets more attention, which may be because it has a solution that's more personally achievable and directly gratifying. Rewiring the electrical grid is not something that you or I could do alone.) We focus on the little things in the hope that they matter, so we can feel like we at least did something when the apocalypse comes. In the aggregate, these little things can matter. But it's not really about using a plastic bottle or not using a plastic bottle (don't think I'm letting you off the hook for your personal habits, though: we used more than 56 billion plastic bottles in the US in 2018[4] and we mostly don't need to), or sipping from a paper straw or a plastic one.* It's much bigger than those things. It's a global problem in the most literal meaning of that word.

---

* On the subject of plastic straws, most of us don't need straws, and yes, the plastic pollution caused by straws is bad for the environment. However, even if the popular estimates are correct and north of several billion straws are lying on the world's beaches, that still would represent only about .03 percent of the 8–13 million tons of plastic that gets into the ocean from land every year. More than half of that plastic

It's about everything we use: what it's made of, how it's made, how we use it, what happens when we throw it away. I hope I can help you understand how complicated this stuff is—if something sounds simple, it probably isn't. There are tradeoffs and consequences for almost everything we buy and use and eat, and if you hear about a policy or a product that sounds like a silver bullet, you're probably not getting the full story.

In this book, I focused on four areas—the Internet and technology, food, fashion, and fuel—that we interact with every day, because, whether you think about it or not, the lives we're all living have something to do with climate change and the environment.

Maybe you don't think it's helpful to hear how big the problem is and how we're making it worse without thinking about it. I agree: the size of the problem and the narrative of personal responsibility is destructive! It makes us feel guilty about everything we do, even though we had no idea and weren't in charge of setting up the cattle industry! It shouldn't be the consumer's responsibility to find out which type of fish is okay to eat, or which inexpensive cashmere sweater is okay to buy (which is not to say you should eat fish and wear cheap cashmere with abandon). Instead, it should be up to the company to produce cashmere responsibly or not to catch and sell fish that shouldn't be caught and sold, since the companies making money from these activities are the experts (theoretically) who control how the product is made. That's a change that we can demand companies make. We

---

comes from mismanaged waste from five countries—China, Indonesia, Philippines, Vietnam, and Sri Lanka. So improved municipal garbage collection would do a lot more for the turtles than a paper straw (and wouldn't risk the moral hazard of people feeling good about not using a straw so they backslide on other plastic-related behaviors), not to mention that I bet your coffee shop still offers to-go cups, which are often lined with plastic, plastic lids, stirrers, creamer pods, etc. So we shouldn't be congratulating ourselves on the paper straw. It might be time to reconsider take-out and our "on-the-go" culture instead.

don't have to buy their products if they are unwilling to at least tell us where they came from.

It may sound cheesy, but as I went through the five stages of environmental grief—denial, anger, trying to use less plastic, depression, determination—while writing this book, I came to realize in a new and powerful way that, in the end, we're not powerless. In this country, we can vote. And that can work. In 1969, pollution in the Cuyahoga River near Cleveland, Ohio, caught fire for the thirteenth time because of the oil and industrial waste that were being dumped into the river, and there was also a massive oil spill in Santa Barbara, California. The next year, activists and politicians organized the first Earth Day, which brought 20 million Americans to the metaphorical and physical streets. One of their goals was to get people to base their vote on one issue: the environment. In the 1970 election, some of those same activists, part of Environmental Action, targeted twelve members of Congress with the worst environmental voting records, nicknaming them the "Dirty Dozen." When seven of the twelve lost, the impact went way beyond those seven elections. It sent a message to all the other lawmakers and led directly to the passage of the Clean Air and Clean Water Acts, two of the most consequential and effective pieces of environmental legislation in history.[5] And it is starting to happen again: as of this writing, after the 2018 midterm election, several newly elected members of the House of Representatives, many of them young women of color, along with some senators, made the passage of a Green New Deal, an actual set of policies to combat climate change, among their first priorities.[6]

No matter what happens, we are going to have to continue to fight to make a difference. The changes we need are big and complicated, and a lot of special interests are aligned against progress, so these new elected officials might not be able to make them happen right away—we might not change the entire electrical grid in one session of Congress—but that's no reason not to start. Making even small changes will definitely be much better than where we are. And better might not be perfect, but better is good.

As citizens, we have a responsibility to put environmentally progressive leaders in office. But it doesn't—it can't—end there. We have to know enough to make sense of what they're offering, to know if it's actually what needs to get done, and to hold them accountable for their actions. If we want to have clean electricity, grow food, manufacture goods, and get around responsibly, we have to understand what it will take to get there and we have to make it happen. Creating a context to understand those issues is what I've tried to do in this book, mainly by talking about our stuff. It's up to us to create a country that takes seriously its obligations to the planet, to each other, and to the people who will be born into a world that looks different than ours has for the last 10,000 years or so. If we aren't paying attention, others with destructive intentions or different motivations might make the decisions for us.

Essentially, what I'm describing is hard work with possibly limited success for the rest of your life. But we have to do it, and at least we will have the satisfaction of knowing we made things better. I hope this book can help, because if we don't get started now, the world ends and the rats take over.

Come on, it will be fun (?).

Technology and the Internet

People often seem surprised to hear that the Internet has any kind of environmental impact at all. I have a feeling that's because the inner workings are invisible, but the Internet has a physical infrastructure, made up of wires and cables, servers and routers. It needs electricity to work, and it is switched on and running all the time.

We don't need to understand how the Internet (and even our computers and cell phones) works in order to use it. We know that it works, but since we don't see how it works, we make all kinds of assumptions, like that it's efficient and logical. But it's not.

The Internet wasn't originally designed to do what we now use it for: shopping, watching movies, social networking, ripping each other apart, hacking each other's elections, etc., etc.,—average, everyday stuff. Those capabilities were added on to an existing design that had been cobbled together to enable different possibilities, mostly allowing government officials to talk to each other. Much of the "design" of the Internet, it turns out, happened as a workaround.

On the one hand, the ease of using the Internet has made technology completely accessible, allowing people all over the world to have information that never would have reached them before, or talk to each other, collaborate, discover, create, and find new solutions to old problems. On the other hand, it allows us to be completely separate, at least intellectually, from the things we are using and the processes by which they are made and operate.

And that includes electricity.

The Information and Communication Technologies (ICT) sector, which includes our devices, data centers (where the information is stored), and network transmission, uses about 1 percent of all the electricity generated around the world and contributes somewhere around 2 or 3 percent of all carbon dioxide emissions, just a little more than air travel and about the same as the shipping industry.[1] The ICT sector is projected to reach nearly 21 percent of global electricity demand by 2030, though in the best-case scenario, it only gets to 8 percent. Either way, that's not nothing.[2]

Why does the Internet need electricity? Explaining that will require a quick detour into the history of the Internet, because we need to know a little bit about how it works and where it is, physically, to understand. Once we know that, we'll have a better sense of the importance of data centers and cloud computing, and you can go from being someone who pretends to understand what those things are to someone who actually does. I want to explore why data centers use as much energy as they do (a lot of which has to do with our behavior and what we want to use the Internet for). Understanding how the Internet works and the different things it has made possible will also help to explain the energy and resource use around online shopping, and if it's better or worse for the environment than going to a store. And cryptocurrency will be explored because it's 2019 and if I didn't write about cryptocurrency, everyone would ask me why, but also because it's interesting. However, I do think the conversation around cryptocurrencies and their energy use has been distracting, in my humble-and-slightly-nervous-that-cryptocurrency-enthusiasts-will-come-after-me opinion.

And it's not just the Internet that has an environmental impact. Your computer, cell phone, and other devices all take a significant toll on the environment. You may have read articles about labor practices in factories where computers and phones are made, and the problems these factories cause for workers, which is important. But the impact is much larger than that: the resources used to make these devices are extracted from the earth in ways that can cause significant damage; toxic chemicals, especially cleaning agents, are used in production and can get into the environment and create health problems for workers and people living around factories or former production sites. And we use the devices in wasteful ways: about three out of four of our devices use electricity when they're off or not being used.[3] Then there's what happens to our devices when we are done with them. Electronic devices are supposed to be recycled, but around the world, as much as 90 percent of electronic waste is improperly recycled or disposed of.[4] Even if you "recycle" your old computer, that might not get done properly. Instead, it might sit in a landfill, where it can leach toxic chemicals. Or it could be (often illegally) shipped to developing countries where it may be "informally recycled": smashed, burned, or otherwise destroyed. (FYI, burning plastic is not good for people or the planet.) The precious metals it contains may be taken out by hand and with significant health and environmental consequences.

Just because you keep me honest: I wrote this book on a computer and did a lot of my research on the Internet, so we're all in this together. We're going to learn more and understand more and hopefully make some better choices together. Join me as I thrust my qualms and anxieties about modern technology (not privacy-related like everyone else's!) onto you.

The Physical Internet

Follow a string of telephone poles for long enough and you will find the Internet. You don't really need to know what you're looking for—a sign will tell you that you're there. It will be somewhere on a fiber cable route, and it will instruct you not to dig anywhere near where you're standing and possibly indicate something about communications and the United States.

I knew that these fiber cable routes and signs existed because of the research I had done for this book, but I was still surprised when I saw one in the wild.

I happened to be cross-country skiing by myself at the time (an activity that my skill level suggests I should not do alone) and I fell. This was in December 2017 in west-central Colorado, when the mountains had seen very little snow. There was about an inch and a half of hard pack snow (aka ice), and the mountains were brown, their scrubby trees poking scraggily out of the soil. The track where I happened to be skiing was a white streak through the desert, not the snowy winter wonderland I had imagined when I was tricked into going skiing.

The point of all of this is that if it had snowed more—if there were more than an inch of snow on the ground—it might not have hurt as much when I fell, and I may not have decided to lie in the "snow" for ten minutes. And if I hadn't been lying there, I might not have noticed the sign pointing out the Internet, which, if it had snowed more by the end of December, would have been buried in the snowdrift that the Rocky Mountain region has come to expect from midwinter.

(By January 4, 2018, when I started writing this section of the book, about a week after this fun excursion, the Colorado statewide snowpack was about half of what it normally is by that date. Warm temperatures—a monthly average temperature of 45.5 degrees Fahrenheit, 7 degrees above normal[1]—had also caused whatever snow had fallen by then to melt. The volume of water in the snowpack, which represents an important drinking water supply for much of the West, is projected to decline by up to 60 percent in the next thirty years[2]—on top of the 20 percent we've already lost since 1915—depending on how much we reduce our greenhouse gas emissions.[3])

Instead, I saw signs proclaiming the presence of the Internet poking out of a trail, winding hundreds of feet above a river, beneath a proud parade of telephone poles, leashed by wires, that stretched into the desert.

There are a few clues that this sign means the Internet is here (it doesn't say THIS IS THE INTERNET on it, unfortunately). The words "fiber cable route" are one clue; that it belongs to US West Communications is another. (US West Communications doesn't exist anymore—it was bought by Qwest Communications International in 2000,[4] and that company merged with CenturyLink, a global telecommunications company based in Louisiana, in 2011.)[5]

But the biggest clue of all is the sign's position beneath the telephone lines. In the United States, most fiber cable routes, which carry both Internet and television signals, follow telephone lines. These telephone lines, which stretch out in a webbed, cross-country network, for the most part, aren't random—many follow the routes of telegraph wires that preceded them. Telegraph wires followed the railroad lines and vice versa.[6] Sometimes they follow highways, which are another important physical network, but railroads were the first formal and government-subsidized network, so that's why I'm focusing on them.

Railroad lines, unlike the highways, were not planned strategically—they are haphazard and were cobbled together by competing companies to connect places and people that, at the time, did not necessarily need to be connected. Railroads often followed the path of least resistance—routes that offered the most even grade across the country, rather than those that made the most sense. That's why Council Bluffs, Iowa, became Union Pacific's starting point for the first "transcontinental" railroad—the path out of there followed a uniform grade along the 42nd parallel across the Great Plains.[7] Since the federal government was subsidizing the land purchases and securing the rights of way for the railroad companies, it was relatively irrelevant which routes they took or how many paths from St. Louis to San Francisco were created, since money was no object. Railroads literally shaped the West as we know it. There are cities that sprung up in one place or another because that was where the railroad was—like Billings, Montana, and Spokane, Washington. Railroads collapsed space and time, allowing people and goods and information to travel farther and faster, realigning people's conceptions of what was feasible and what was convenient (much like the Internet). But most of all this was a physical demonstration of American expansionism by giving white settlers easy access to the West, and of Gilded Age excess (since many of the railroads that were built were unnecessary). The construction of the railroad was a corrupt process, generating wealth for the few at the expense of the many, which we call capitalism. (These big ideas about railroads largely come from historian Richard White's incredible book *Railroaded: The Transcontinentals and the Making of Modern America*.)

When telegraph lines were beginning to be installed across

the country in the nineteenth century, telegraph companies realized it would be more efficient to install their wires alongside railroad tracks, in part because they took the easiest routes across the country, but more importantly because it was easier to secure easements from one entity—a railroad company—than piecemeal landholders in every state from New York to California. Railroad companies were only too happy to have the telegraph cables lining their routes—they generated passive income for the companies, and they helped station managers communicate better about which trains were where at any given time, helping them to avoid accidents.[8] As westward expansion continued, the telegraph lines sometimes preceded the railroad construction, which allowed the companies to communicate with the outposts of the possible network, places where they were considering laying down track.

When telephones replaced telegrams, telephone companies used and expanded the paths that had already been worn across the continent. Once the interstate highway system was built, telephone companies (which also installed cable for television) did the same thing there, too.

And the railroad companies benefited in other ways, adapting to a changing transportation and communication situation. In 1972, the Southern Pacific Railroad, which previously operated telegraph wires along its tracks, decided to use the existing communication lines for long-distance dialing. By the middle of the decade, it was selling time on its private microwave communication lines to individual customers. That became Sprint, which was an acronym for Southern Pacific Railroad Internal Network Telecommunications. In 1988, Philip Anschutz, an American businessman and owner of Coachella, bought the Southern Pacific Railroad company and negotiated the right to lay fiber along the SP tracks (and those of other railroad companies). With another of his companies, Qwest Communications, he began installing fiber and switches for the company's own use. In 2000, he bought US West Communications, the owner of my sign, the one I had seen while sprawled on the ice-covered ground in a fit of self-pity/loathing.

I tried to find out who was responsible for laying this exact piece of cable. I couldn't tell if it was US West Communications itself, US West when it was part of AT&T, or MCI, which had had a contract to lay fiber cable from the government to set up the Internet in its early days, Qwest on behalf of MCI, or someone else entirely. I asked CenturyLink, which bought Qwest in 2011. They said they couldn't tell me who laid the cable or who even owns it. I sent a picture of the sign; they couldn't say if they had sold it and the sign had just never been changed. I gave them the approximate latitude and longitude of this sign. They said they could be of no further assistance.[9]

But anyway, the point is that once the fiber cable routes, which now carried cable television signals, existed, they could also carry Internet signals. During the economic explosion of the ICT sector in the late 1990s, telecommunications companies raced to lay down cable alongside railroad tracks. Other traditional utilities—natural gas, electricity—offered their existing networks as a skeleton for fiber cable pathways, another physical manifestation of the dot-com boom.[10]

We can see the benefits of being near old railroad termini or network hotspots: Google, Microsoft, and Facebook have all built data centers in Iowa over the last decade, and Iowa—specifically, the city of Council Bluffs—was where the first transcontinental railroad began. It's also worth noting that Iowa's electricity prices are slightly below the national average, and the state also doesn't levy sales tax on electricity usage.[11]

This may seem like a long digression that doesn't seem to have a lot to do with the environment, but believe me when I say I have a point: as much as the Internet represents the exchange of ideas and information, it's also a physical thing, a network that connects us, materially, in cables and routers and blinking lights all over the world.

Apart from the cables, there are a few other important physical pieces of the Internet. One of them is your device (computer, phone), which uses electricity to operate in general and even more electrical energy to use the Internet specifically. But there are others, some of which may be well known, like data centers or server farms, and others that are less well known, like Internet exchange points (IXPs). These have a spatial environmental impact—they're sometimes physically really big—but more importantly, they require electricity, which usually requires the burning of fossil fuels, and exact a much greater environmental toll than just the electricity the devices use.

When you connect to the Internet, you are actually connecting to a server in your network, which is connecting to the larger Internet beyond your network. These servers also store information, in the form of websites or email clients or other online data platforms. Servers, like all computers, need electricity to perform the tasks we ask of them whenever we want, so they are always on to keep a network connection. The constant processing also means that they get pretty hot, so even more electricity is used to cool them down. Companies, schools, government agencies, and other network-connected environments often have their own servers; others may choose to co-locate, using capacity on equipment owned by someone else, most likely in a remote location, possibly even in the cloud.

The way that information from those servers travels across the Internet is hard (for me) to understand. Not only is it complicated computer science, but the networks are all higgledy-piggledy, built on top of each other, and the Internet is not an efficient, transparent system that lends itself easily to comprehension.

But to put it somewhat simply: Internet exchange points are an important part of the way Internet traffic travels—they connect different networks in a direct, physical way. If you and I use different Internet service providers and I want to send you an email, my message has to leave my network and go to yours. An IXP makes that possible, and does it more quickly than if my message had to go to the larger Internet backbone, which both of our networks are connected to. These IXPs all over the world tend to be congregated around big population centers or the coastal landing sites of transoceanic cables; data centers and other physical pieces of the network also crop up near these exchange points.

Conceiving of the physical existence of the Internet helps us to understand how it works as a system, and the different kinds of resources it requires—primarily, energy and land. While Internet companies pay the monetary cost for their land and electricity, they typically haven't had to pay for the environmental costs associated with electricity generation. The rest of us (especially people who live near coal-fired power plants, who may breathe in the by-products of coal burning or find it in their water supply) do.

This may not seem to have a lot to do with the environment. But if you keep in the back of your mind that the Internet is a physical system, when you hear about the worldwide expansion of Internet access, the increasing frequency with which we are all online, and the proliferation of data, you may also think about the electricity required to make those things happen and imagine the specific places where the Internet is, and how it may change those places.

The logical next questions: Why is the Internet congregated in a few places? And where are they? Why them? Follow-up: Why me?

If a geographic location comes to mind when you think of the Internet, it is probably Silicon Valley. But it might as well be Northern Virginia.

The original Internet was a product of the US Department of Defense. Known at the time (in the 1960s) as ARPANET,* it was intended as an internal communications network that could survive a nuclear attack. As a military creation, ARPANET was closely associated with the Pentagon, which is in Northern Virginia. When the government began work on what would become ARPANET (and later, the Internet), it relied on the services of computer scientists from academic institutions and federally funded government contractors, who moved into an area that was close to Washington, DC, and the Pentagon, where commercial activity was allowed, directly accessible by the relatively new interstate highway system, and with open land. That area, part of Loudoun County, was known as Tysons Corner (also the site of a microwave tower that formed part of an earlier nuclear-attack-resistant communications and underground bunker network, chosen because of its history as a good signaling location during the Civil War) and, along with the surrounding sprawling highway-connected suburbs, it became the country's Internet capital, home to defense contractors and IT companies.[12]

The process by which ARPANET became the Internet is pretty complicated and would involve poor explanations by me of things like packet-switching and TCP/IP protocol and government bureaucracy, and neither you nor I would enjoy it. If you want to read more about this history, I suggest *Internet Alley: High Technology in Tysons Corner, 1945–2005* by Paul E. Ceruzzi and Janet Abbate's foundational book, *Inventing the Internet*, part of the Inside Technology textbook series.

Let's just say that the government's initiative and money brought a lot of computing companies to Tysons Corner, and when

---

* ARPA stands for Advanced Research Projects Agency, an agency within the defense department that developed new technologies for military use.

the National Science Foundation took over building out the infrastructure of the Internet, one of the earliest network access points (a precursor to the IXP), known as Metropolitan Area Exchange-East, was dreamt up and built in a basement in Tysons Corner by several tech industry companies.[13] The Internet backbone (the main high-speed fiber-optic cable network that connects large and important networks, routers, access and IXPs) grew from this humid, low-lying swampland. As Ingrid Burrington wrote in her masterful series in the *Atlantic*, "Networks build atop networks, and the presence of this backbone in Tysons Corner led to more backbone, more tech companies, and more data centers," highly concentrated in this one spot.[14] The Loudoun County government estimates that 70 percent of global Internet traffic passes through there,[15] a result of the many historical contingencies that brought the Internet to Northern Virginia. Apple's first retail store wasn't in Cupertino—it was in Tysons Corner.[16] (Its second store opened three hours later in Glendale, in Southern California.)[17]

So what does the history of the Internet's physical existence show? To me, it demonstrates that no one anticipated how useful the Internet would be or that it would transform every aspect of modern life. Proof: the busiest Internet exchange in the world was created in a non-descript building in suburban Virginia.

The Internet was also built where it was because there was cheap open land, and the consequences of paving it over and filling it with office buildings appeared minimal at the time. There was and continues to be cheap energy, largely provided by fossil fuels. At the time that the Internet was evolving, even more of the power it ran on came from coal, and the relative inexpensiveness of the electricity probably allowed it to grow. And, it's hot and humid there, so the servers get even hotter, and having available cheap energy became even more important.

The pace of the Internet's growth continues to astound: just about every two years, media companies and industry groups report breathlessly that 90 percent of all data on the Internet was created in the last two years.[18] (Translation: the growth is exponential.) We are also using the Internet differently than we used to, in ways that are increasingly more energy-intensive. Video streaming, as a percentage of Internet use, just keeps growing: in 2010 it was 40 percent;[19] by 2015, around 70 percent; by 2020, it's expected to reach 82 percent.[20] Around the world, more than 4 billion people use the Internet, and most of them are watching video, the quality of which keeps improving, requiring more data transfer and therefore more electricity.

And more and more people are coming online. In 2018, 88 percent of American adults said they used the Internet, 26 percent said they go online "almost constantly," and 65 percent used broadband Internet connections at home.[21] Among adults, 77 percent had a smartphone, a number that has more than doubled since 2011.[22] In 2018, just over half of all pageviews happened on a smartphone,[23] and more than 60 percent of global mobile phone users were expected to have an Internet-connected phone by 2018.[24]

As the Internet grows in size—measured in bytes—and by the number of people who use it and the networks we build, we have to make sure that it grows sustainably and responsibly. Data centers are a good example of why we need to do that, and some of the solutions in that area might provide a good road map to how it might get done.

Bringing the Cloud to Earth

Stretching from the East Coast of the United States across the Atlantic Ocean to continental Europe are more of these fiber-optic cables, possibly with fewer signs indicating ownership, definitely making transatlantic Internet activity possible.

At the Keflavik International Airport, a once (and possibly future) NATO base just outside of Reykjavik, Iceland, submarine cables from Europe rise out of the ocean at the seaward edge of the tarmac and snake into a series of low-slung buildings. For all the hard work it took to get them there and the futuristic, near-fantastical achievement of getting Internet signals to cross the ocean, the cables end up, just like they do everywhere else, in what looks like an office park anywhere else—suburban Virginia, for instance.

These buildings are also home to a high-performance-computing data center, run by a company called Verne Global, which provides hosting for other companies that perform services like car-design modeling and Bitcoin mining.

Verne Global is not located in Iceland because it's an Icelandic company (it's not); it's there to take advantage of the fiber-optic connection and the cheap, (nearly) carbon-neutral geothermal (since geothermal energy produces some carbon dioxide emissions) and hydro-powered electricity in Iceland, 90 percent of which isn't needed for the everyday activities and needs of Icelandic people, so there's a lot to spare for things like data centers and aluminum smelting. Verne Global's founders realized that if they put their servers in a place that is cold for much of the year, they wouldn't have to waste as much energy on keeping them cool and could use the (unpleasant) ambient temperatures instead (I do not like winter). It came as a secondary benefit that, as a result of those things, their data center is more environmentally friendly than those in other places.[1]

And the gamble may pay off: as our data needs grow, we're going to need more servers and therefore more electricity to power them. Requiring less electricity, at a low, consistent cost, with the added benefit of near carbon neutrality, is turning out to be a big advantage for Verne Global, which prominently advertises its sustainability.

Verne Global's data centers, and others like them all over the world, make up the cloud. Calling it "the cloud" is misleading; it is actually a physical, electricity-guzzling system that connects the Internet to earth.

There are other kinds of servers (which use more electricity and make up more of the Internet), but I'm going to concentrate on the cloud because that's where the Internet is headed. The cloud is much more efficient than traditional servers, and it's where Alphabet, Apple, Facebook, Amazon, and Microsoft, five of the biggest companies in the world by market capitalization, store their data. Combined, these companies host sites that constitute more than one-third of the Internet, so the decisions they make in any area have important implications for how other companies behave and how consumers use their products and, by extension, electricity and other resources.

Everything we do on the Internet—a Google search, sending an email, watching a YouTube clip—requires energy. In 2009, Google estimated that a typical search uses as much electricity as it takes to power a sixty-watt light bulb for seventeen seconds and results in the emission of 0.2 grams of carbon dioxide.[2] In 2009, we used the Internet differently than we do now, and we create a lot more data now than we used to: the average web page, mostly because of photos and videos, is now 3 MB, which is larger than the entire PC game Doom from the 1990s.[3]

The massive proliferation of data is due to video, which makes up nearly two-thirds of Internet traffic.[4] The way we watch video online has changed profoundly in less than a decade. In 2011, when viewers streamed 3.2 billion hours of full-length movies and TV shows over the whole year, one study found that video streaming in most cases contributed fewer carbon dioxide emissions than DVD viewing, especially when the process of watching a DVD (which included production and transportation of the DVDs) entailed a car trip to buy or rent a DVD.[5] In this study, data center electricity use accounted for less than 1 percent of the total energy required to stream a video.[6] Plus, streaming helps us avoid making new material waste, in the form of the discs themselves and their plastic cases.

That seems quaint now, when about 19 percent of Internet traffic in North America is spent streaming video just on Netflix;[7] globally, Netflix uses up about 15 percent of Internet bandwidth.[8] In 2018, an estimated 228.8 million people in the United States watched digital video content,[9] and they watched, on average, about eighty-two minutes of it each day.[10] That comes to more than 114 billion hours per year, including YouTube videos and not including the nearly 4 hours of actual TV per day they watched,[11] all of which leads me to believe that most human brains are melted piles of flesh. One hundred and fourteen billion hours per year (fewer if you don't include YouTube videos) is about thirty-five times what we watched in 2011, and almost four times what we watched online and on DVDs that year combined. The study's authors estimated that watching video online resulted in 0.4 kilograms of carbon dioxide emissions per hour, adding up to 1.3 billion kilograms per year.[12] Assuming those emissions per hour are still about the same (given that some data centers will have aged as new ones are brought online), that corresponds[13] to 45.6 billion kilograms or 50.3 million tons of carbon dioxide every year, just from Internet video. (The US emits 6.511 billion tons of carbon dioxide equivalent each year,[14] so Internet video streaming makes up about 0.77 percent of emissions, which is not a lot, except that it's just Internet video.) Plus, about 30 percent of Americans still buy and rent DVDs (Schlossberg's Believe It or Not!), which is a significant drop from 2011, but not zero.[15]

It's possible that the savings from producing fewer DVDs mean we're emitting less overall, and we're probably creating less waste, but we're watching so much more video than we used to that the savings may not be enough to cancel out all the energy we're using by streaming hours of *The Great British Bake Off* over and over and over again until our brains turn to slime and we die.

People who study energy and efficiency call this phenomenon the rebound effect: when savings from efficiency or dematerialization are canceled out by corresponding growth of use.

There are two ways to ensure that the Internet, which will inevitably get bigger as more and more services move online, grows responsibly and sustainably, especially with the advent of cryptocurrency, artificial intelligence, and self-driving cars, among other services that we probably can't even imagine yet.

One is to operate these data centers efficiently, and that's something that big Internet and data companies have worked at. Cloud storage and hyperscale data centers have allowed the growth of energy use to proceed relatively modestly. Without the efficiency improvements, especially in cloud centers and elsewhere, the electricity use in data centers would have doubled from 2010 to 2014, but because of the gains in efficiency, it remained essentially flat, accounting for about 1.8 percent of electricity use in the US.[16] How have they done that? Servers in these companies' data centers operate at greater capacity (there are fewer idle servers, and the servers use more of their capacity), so they manage temperature better, and they use recycled water or wastewater for cooling.

And scientists who study data center energy use and energy efficiency have a lot of faith in these companies to maintain the efficiency of their systems. Eric Masanet, an associate professor of mechanical engineering at Northwestern University who has studied data center energy efficiency for the Lawrence Berkeley National Laboratory, the Natural Resources Defense Council, and the International Energy Agency, sees positive trends: the increasing efficiency of our devices, mostly a result of trying to extend battery life; data center companies' preoccupation with efficiency, reporting energy use, and using renewable energy, in part because of the negative publicity they have received; and network transmission becoming more efficient with every generation of cell phone.

"Every couple of years the energy use of data gets cut in half," Masanet told me. "How long can those trends continue?"[17] So far, the improvements in efficiency have been "enough to offset the amount of data we're consuming as a society, and have been improving much more rapidly than any other technology out there." And the numbers back Masanet up. Over the last decade, data centers have become about 1.5 times more energy efficient, on average (and about 2.25 times more efficient if you are Google).[18] But there is still room for improvement. Some studies suggest that even though data centers have become much more efficient than they were originally projected to be, they could still cut their energy levels by about one-third, just by making small changes.[19]

Not everyone is sold on this narrative. Adam Nethersole, a Verne Global executive, said, "The amount of power that we'll need to fill our data requirements—because of the boom of autonomous vehicles, artificial intelligence, et cetera—is going to hit a point where the availability of the power to do that is going to rapidly diminish."[20]

For Nethersole and others, what matters more is where the energy comes from. For Verne Global and other companies, many of which are in Scandinavia, the carbon-neutral or low-carbon energy sources make the biggest difference. And some companies are on board with that idea: Google has data centers in Finland;[21] Facebook built some in Denmark[22] and Sweden.[23] As I wrote above about Verne Global, placing data centers in these cold climates means that they don't have to use as much electricity to cool the servers, which could mean that they use less energy overall.

And there are certain applications in which using less electricity or using carbon-neutral electricity can make an even bigger difference. With that said, I'm going to talk about Bitcoin and cryptocurrency. If you don't understand what Bitcoin is, you are not alone, because most of the people who say they understand what Bitcoin is are just pretending. Some people—the haters, out there, generally—might ask why this entire book is not about cryptocurrency and Bitcoin, and they're right. The rest of the book will be spent explaining what Bitcoin is. Kidding! But I am interested in cryptocurrency and how it might continue to function and develop.

Anyway, Bitcoin is a prominent cryptocurrency, a word used to describe digital assets (currencies) that are exchanged online, and are cryptographically secured, often by blockchain technology, which is essentially a public ledger of all transactions. In general, cryptocurrencies exist without a centralized bank or institution, the idea being that decentralization, the public ledger, and the code make these assets more secure. The word *Bitcoin* actually refers to two things: Bitcoin the unit of code, which represents the digital asset itself, and Bitcoin the network, which maintains the ledger of all the transactions involving Bitcoin the unit of code. Here's (sort of) how Bitcoin works: there are a finite number of Bitcoins, and the supply is controlled by the algorithm that underpins the whole system. A small number of Bitcoins will be released every hour (with a smaller and smaller amount released each time), until all 21 million Bitcoins have been birthed into the world. The way that you get a Bitcoin is by "mining" it. A Bitcoin is mined by having a computer run software that solves complex mathematical problems whose answers are processed and then added to blocks of validated Bitcoin transactions, aka blockchains. We did it. We got through an explanation of Bitcoin together.

Mining Bitcoin—because it requires computers to be on, running software, and performing complicated mathematical tasks that people cannot do—uses a lot of electricity, which is why we are talking about Bitcoin.[24]

Bitcoin and other blockchain-based cryptocurrency mining has taken off in Iceland for the same reason that attracted Verne Global, which does host some Bitcoin mining: reliable energy at a cheap, fixed rate. The amount of energy that Bitcoin and other blockchain technology uses is not small, but the way that people talk about it, you would think it was the biggest energy crisis we're facing.

First, there is more than just Bitcoin, and other blockchain technologies (like Ethereum) are trying to switch over to a less electricity-intensive way to ensure security, called proof of stake, while Bitcoin operates as proof of work,[25] but neither the Ethereum community nor any other major player in cryptocurrency has fully made the switch yet. It's not especially salient for our purposes to understand the difference between proof of work and proof of stake. The important takeaway is that proof of stake has energy-saving potential that could be realized over the next three to five years, depending on whether it actually works and whether it's adopted on a big enough scale to make a difference.[26] Ethereum is projected to grow, but as of this writing, Ethereum's market share is about one-sixth of Bitcoin's, far and away the biggest cryptocurrency.[27]

Second, there is a lot of disagreement about how much electricity Bitcoin mining actually uses. Some people are incredibly concerned: a paper published in May 2018 argued that Bitcoin would use 0.5 percent of global electricity by the end of the year.[28] Analysis based on that study found that a single Bitcoin transaction uses, on average, 1,005 kilowatt-hours, while 100,000 Visa transactions use 169 kilowatt-hours.[29] Others have found that study to be problematic, arguing that it used shaky methodology[30] and that Bitcoin mining doesn't use anywhere near that much electricity. Either way, we can say that Bitcoin mining definitely requires a significant amount of electricity.

Clearly, Bitcoin and other blockchain-based currencies could be a real problem depending on how they develop. Some people try to argue that they will save energy, given that they use fewer resources—physical, personal—than global banking. But that argument presumes that we will only use Bitcoin, which is not happening, and I'm not sure that even the biggest Bitcoin or blockchain enthusiasts predict that happening anytime soon. A side note: while credit cards may require less electricity per individual transaction, the global banking system still uses electricity associated with the Internet, and that's just scratching the surface of the resources it uses—metals, linen (for the bills), paper, plastic, pens on chains, and electricity to power its ATMs and physical bank buildings, whose lights are kept on 24 hours a day all over the world.

In essence, that explains Bitcoin, and now we are done talking about it forever.

But the problem that cryptocurrencies encapsulate is the

way we use data more generally. Our data needs require significant amounts of energy already, and the way we behave online is progressively and continuously more data-intensive, outpacing our ability to store it. And we treat data as if it's infinite, and things like cell phone plans with unlimited data reinforce that perception.

So we could change our attitude toward data, which seems unlikely at the rate we're going, we could increase efficiency, which is happening but may ultimately reach a limit, or we could change the energy sources used to generate electricity if we want to grow the Internet sustainably, which is another approach that data hosts have recognized. Plus, using renewable energy is good for business: costs are lower and public relations benefits are enormous.

There are basically two ways to power the Internet renewably (though this isn't specific to data center and tech companies). First are Renewable Energy Certificates (RECs), which historically have been popular, including among data center companies. RECs are registered certificates that allow the purchaser to claim that they used renewable energy, essentially by paying for it to exist. For each megawatt of renewable power, one REC, which is independently verified and tracked, is created. Most renewable power projects send their electricity to the same grid as fossil fuel powered generators. When a customer buys power from that grid, they're getting power generated both by renewable energy and fossil fuels. Essentially, by buying an REC, the purchaser can claim that they're only paying for the renewably generated electricity. In this case, data center companies buy RECs from the power companies. Some experts argue that RECs can't necessarily displace power that comes from fossil fuels, mostly because the prices are set too low to allow renewable energy companies to build additional capacity. Others argue that it's a good way to displace fossil fuel emissions and support clean energy.

But wait, there's a better way! Some companies, again, including tech companies, have also entered into agreements with energy companies to make sure that renewable energy capacity will exist on their grid, called a power purchase agreement. They might pay a utility specifically to install additional renewable energy sources, rather than just paying for existing ones. Google signed a deal in 2017 to power one of its European data centers with electricity from one of the largest solar farms in the Netherlands.[31] And, taking it a step further, some companies have built their own renewable energy facilities near their data centers or on the same grid. Many do both.

Still, some scientists and environmental advocates are skeptical about many of the renewable claims that many tech companies make. They worry about where data centers are located, how much of the electricity the renewable energy is supplying, and what happens when that's not enough. For instance, if there is a spike in electricity demand or the sun isn't shining or the wind isn't blowing, data centers still need electricity, and where in the country you are dictates the source that electricity comes from. In Oregon and Washington, for instance, there's a good chance it's coming from hydropower. But it depends on the state's policies, what the energy mix is, and what contingencies it relies on when cleaner fuels can't meet the demand.

When it comes to siting data centers, Amazon provides a useful case study. Amazon Web Services (the data center arm of the company) sells data storage to other companies, and, according to a 2012 estimate, about one-third of all Internet users visit sites hosted by AWS each day.[32] Netflix, the *New York Times*, Reddit, Vimeo, the *Guardian*, and the *Washington Post* are just some of the sites that use AWS to host some or all of their data.[33]

In the last few years, AWS has significantly invested in renewable energy, completing or announcing the construction of six solar farms and four wind farms, and they claim to have gotten half of their energy from renewables by January 2018.[34] However, they do not release information on their carbon footprint, so we can't independently judge the percentage of energy they got from renewables. (I asked representatives from AWS for these numbers several times over the course of eight months, but eventually got my information from the AWS Sustainability website. They were very good at asking me questions, which I answered, and either not really answering mine or mostly providing me with information already available on their website. It was impressive and I salute them.)

But the states where many of their data centers (and lots of others) are—Ohio and Virginia—predominantly use fossil fuels, including a (un)healthy amount of coal. In these states, two really big and powerful utility companies control most of the power plants, and more importantly, they decide where the electricity comes from. Since these states don't have policies that require a certain amount of electricity to come from renewable energy and because of where these states are, less than 5 percent of the electricity comes from renewables, and about 30 percent is from coal.*

What's less good is how the data centers make sure they never lose power. The grid and other electricity infrastructure—power plants, transmission wires, substations, and more—often suffer damage for a variety of reasons, including old age, weather, encounters with wildlife (good/adequate name for a TV show if anyone is asking), and human-caused damage (a car crashing

---

* These states have pretty lame renewable energy targets, which they aren't even meeting. In Ohio in 2017, coal provided 58 percent of the electricity, 24 percent came from natural gas, and 15 percent from nuclear power. Just 3 percent came from renewables, below the state's goal of 3.5 percent for that year. In 2017 in Virginia, 50 percent of the electricity came from natural gas, 12 percent came from coal, 33 percent from nuclear, and 6 percent from renewables, including biomass and hydropower. Nationally, coal provides about 30.1 percent (as of October 2018) of all electricity, and renewables provided 17.1 percent—on average, absolutely smoking both Ohio and Virginia.

into a telephone pole, for instance). Because of the imperfections of the electricity system, many data centers have backup generators, many of which are powered by the most reliable and cheapest energy sources: dirty fossil fuels, usually diesel. If you have ever been around diesel, you may know that it's not super pleasant when burned: it releases particulates and other pollutants harmful to human health. (Representatives from Amazon Web Services would not tell me what kind of backup systems they use, though they assured me they do have some kind of backup system.)

Microsoft operates thirty-seven diesel generators at their data center in Quincy, Washington. When citizens of Quincy formed a group called Microsoft-Yes; Toxic Air Pollution-No, which challenged the state's decision to grant Microsoft permits for all of those generators, they were unsuccessful.[35] Yahoo, Dell, and Sabey also operate data centers (with diesel generators) in the town, too. In August 2018, Microsoft applied for a permit for seventy-two generators at the site.[36]

In Silicon Valley, diesel engines (including those powering generators) at around 2,500 locations emitted toxic amounts of particulate matter (amounts above the "trigger level," which are expected to cause or significantly contribute to adverse health effects). AT&T was the greatest single emitter in the Bay Area, releasing about 670 pounds of particulate matter in 2015.[37] Apple, Microsoft, Google, Verizon, Equinix (an IXP operator), and Sprint feature in the top one hundred polluters of this kind of pollution, which is impressive, given that so do some airports, the Union Pacific Railroad, the East Bay Municipal Utility District, Travis Air Force Base, and several hospitals, which also need backup generators. That's not to say that those places should have more license to pollute, but that they perform essential services (transportation, healthcare, public safety–related activities), some of which, at present, can't be performed without diesel fuel.[38]

With these generators in operation, it's hard to be convinced that data centers are "100 percent renewably powered" or carbon neutral. Plus, with many of these companies, it's unclear what carbon neutrality actually means: whether the electricity to power the data centers comes from carbon-neutral (but nonrenewable) sources, like geothermal and hydropower, or whether the associated emissions are offset and how.

There's not a villain here, and I don't believe (or try not to believe) that some of the companies named and described here are cynical enough about their customers that they want to both dupe us and destroy the planet—or at least hamper its chances for survival—at the same time. I just think that the system as a whole wasn't engineered to work as a coherent system, and that's part of the problem.

I also don't want to overstate the problem created by the Internet—it's not the biggest source of carbon dioxide emissions or the biggest consumer of electricity. Internet and tech companies have big incentives to improve their efficiency—electricity is a large expenditure, so reducing their needs means less overhead. More efficient servers also means fewer servers, and doing well by the environment aligns with other aspects of their mission (or at least their branding) of saving the world.

And it's not as if the Internet and technology are alone in their hidden or unaccounted for or not widely known environmental impacts. But the ICT sector is an important and growing part of our world, so we need to pay attention to how it develops, and without understanding how it works, I don't think that's possible. If environmental advocacy organizations are the only ones asking questions and making demands (Greenpeace and the Natural Resources Defense Council do a good job of this), we're abdicating a lot of power and responsibility that we have as consumers of Internet data and customers of Internet companies to make changes.

If you feel like I'm giving you mixed messages on how good or bad the Internet is for the environment, well, sometimes you have to live with uncertainty. It's complicated! It varies based on where you live and what you do online. It's contingent on policies, regulations, and technological developments. The Internet is sometimes good and sometimes bad. Throughout its history, the Internet has been changed and developed and improved by its users, and that includes us, so it can include our values.

That should also mean that we are responsible in the ways that we apply it to the physical world of goods and materials. I'm talkin' 'bout e-commerce.

Taking It Offline: E-Commerce

For the record, UPS trucks actually do make left turns, though rarely. You may have heard differently if you've ever read a news story about UPS, or maybe you thought it was FedEx. Either way, it's not true. UPS trucks mostly don't turn left—maybe one out of every 10 turns—and they haven't for several decades. The drivers do, however, try to make their deliveries on the right-hand side of the street.[1]

At some point, a UPS truck will have to turn left. It's only natural. But turning left at a major thoroughfare, which the company tries to avoid, a UPS spokesman told me, means that you might have to wait for the light to cycle to make that turn,[2] wasting time and fuel, not to mention spitting out carbon dioxide and other polluting gases, or harmful particles. If you're turning right, you can just go for it.

UPS is the biggest transportation company in the world in terms of sales, with a fleet of trucks, planes, and bicycles and a freight logistics network that uses ships and railroads and jetpacks and hovercrafts and Razor scooters.[3] (Spot the lies!) In the US, UPS accounts for about 57 percent of the parcel delivery market;[4] globally it's around 22 percent.[5] What UPS does matters, especially as the worldwide freight logistics industry (aka shipping packages) continues to grow, and much of its growth is now coming from a new place: the Internet.

In 2017, UPS estimated that half of all the packages shipped across the US were business-to-consumer deliveries.[6] More than two-thirds of Americans have ordered something online,[7] accounting for about 9 percent of all retail in the US, and adding up to around $474 billion. Globally it's a little more—10.2 percent[8]—though it varies from country to country: in China, about 16 percent of all shopping took place online in 2015,[9] but it is projected to more than double by the end of this year; in Japan, e-commerce represented about 6.7 percent, and is only projected to grow by 3 percent through the end of 2019.[10] The percentage of shopping that takes place online has doubled since 2011 and is projected to grow to about 18 percent of all retail by 2020.[11]

But given the way we talk about e-commerce, you might think it was the only way that anyone bought anything. This isn't to diminish the important place e-commerce has and how widespread it is, but it seems like a topic that gets the sign-of-the-times, this-is-so-wasteful treatment that few other areas of modern consumption do.

If we leave the lights on all day, we don't have to deal directly with the by-products of burning fossil fuels, or we don't feel like we do. But the cardboard now comes into our house and piles up, and it didn't used to be there. A decade ago, ordering something online felt like a big deal (every day was Christmas!). Now, we order things from all over the world as a matter of routine, and we can reasonably expect that they will arrive tomorrow, or the next day, or maybe even in a few hours. Something this convenient, which has not always existed within my lifetime, couldn't possibly be good for the environment, right?

Surprise! It's not that simple, and it's actually pretty difficult to calculate the environmental impact of e-commerce. Everyone counts a different way. Do we count the electricity use associated with the online shopping? And the embodied emissions of the product itself? And how do we count how it affects personal travel? Do we assume that every shipment means one fewer trip to the store? But what if on that trip you had stopped in at the store on your way home from work, a trip you were making anyway, or you went and bought a few things at a few different stores on the same trip? How much more cardboard is e-commerce causing us to use? It's complicated!

The easiest place to begin is the cardboard. Even as e-commerce has grown, cardboard production in the US hasn't kept up. In the last ten years, the American corrugated industry (which makes cardboard for packaging) has grown by somewhere between 1 and 2 percent overall, while e-commerce has almost quadrupled (in terms of money spent online). That's not to say that there's not that much cardboard—386 billion square feet of cardboard was made in 2017, growing by almost 3 percent compared to 2016—but we produce less of it than we used to. In 1999, when the industry was at its peak, the cash flowed like water, and the corrugated industry was awash in sex, drugs, and rock and roll (I presume), it produced 405 billion square feet of material.[12]

But this is, at least in part, a question of substitution. Amazon, for instance, ships products in their original packaging so they don't have to be repackaged. Still, a representative from the Fibre Box Association (which represents corrugated cardboard producers in the US) can't say if more packaging is being used for e-commerce specifically than had been used for shipping to retail stores. So we're not using more cardboard than we used to, though individually, we may be more directly using the cardboard. The Fibre Box Association representative told me that items are packaged more efficiently now, so we may use less cardboard per shipment, which helps account for why production in the US has remained relatively constant.

Around the world, cardboard production in the largest-producing countries has mostly leveled off over the last few years, but it has grown rapidly in Asia. Historically, manufacturers in

Asian countries used wood to package their products because it was cheap and durable, but they have been increasingly switching to cardboard; the demand in Asia is expected to grow by 14 percent over the next three years.[13]

Producing cardboard is a very energy-intensive process: it often requires recycling cardboard by breaking it down and then making it back into cardboard, which uses a lot of energy; or else processing lumber into wood pulp and making that into paper, which is then crimped and glued together to make it into corrugated cardboard. Making all paper is the third largest industrial use of energy in the US, after chemical production and petroleum refining.[14] About 2.2 percent of all the energy consumption in the US is spent making paper.[15] Plus, cardboard is heavy, so it also takes a good amount of energy to move it around.

We're relatively good at recycling cardboard, compared to other recyclable materials, but that doesn't mean we're anywhere near good enough at it. According to the FBA, cardboard is the most widely recycled packaging material in the US. In 2013, almost 90 percent of cardboard produced in the US was at least collected for recycling,[16] and 48 percent of cardboard is made from previously recycled material.[17] If you compare that to plastic packaging, it's staggering: in the US, only one-third of plastic bottles are collected for recycling;[18] around the world, only 14 percent of plastic packaging is recycled.[19]

However, we're getting worse at recycling it. In 2017, 300,000 fewer tons of corrugated containers were recycled than the year before, even though domestic consumption grew by 3.5 percent. Because of e-commerce, more cardboard is coming directly to the customer, instead of going to a retail store. Consumers are worse at recycling than retailers, who can flatten their boxes, bale them together, and have a financial incentive to resell them. As consumers, we don't do that at all, or we do it in smaller numbers.

Retailers recycled around 90 to 100 percent of their cardboard; consumers top out at closer to 25 percent.[20] And if not enough cardboard is being returned to be recycled, producers will have to go back to the source to make more cardboard. And that means cutting down trees.

What else can we say about e-commerce? Glad you asked. Most studies about e-commerce focus on the transportation-related impacts, but there are not a lot of definitive conclusions about what those impacts are. One study that looked at traffic patterns in Newark, Delaware, over time showed that trucks, as a percentage of total traffic, had remained the same from 2001 to 2010, despite the growth of e-commerce over that period. But there were more delays in traffic, meaning that cars and trucks were in transit for longer—perhaps because trucks stopped periodically to drop off the packages, possibly causing traffic and using fuel and emitting greenhouse gases and soot particles while not moving.[21]

Another study found that while trucks account for about 7 percent of all traffic, they are responsible for 17 percent of what experts consider to be the costs of traffic: more fuel purchased and more time spent in the vehicle, adding up to $28 billion annually.[22] Instead of dropping off at businesses, which are often on bigger roads, trucks now have to go to residential addresses, causing congestion where there didn't used to be as much. There's also the problem of failed deliveries—when no one is home to receive the package—which may be even more annoying for our national freight logistics system than it is for you, because it could mean another trip for the truck or more time on the road.

And when we buy online, we return more. Some studies have found that we return about 35 percent of what we buy online, compared to between 6 and 15 percent when we buy in an actual store.[23] The returns might mean that the trucks have to come back to pick up our package and return it to wherever we bought it from, most likely cheapsocks.com, a company I have invented. Or you might have to drive to the store or to a shipping service office or drop-off yourself. About half of what we return doesn't get taken back into the store or company's stock—it just gets thrown in the trash, most likely ending up in a landfill.[24]

Trucking can actually be more efficient than personal car travel—trucks carry lots of different packages and make multiple stops per trip. The average American drives about 7 miles on each shopping trip.[25] Other studies have suggested that the increased traffic might not be a result of more trucks in our cities and towns. Instead, it may be that people who used to spend time driving to the store to go shopping are now driving somewhere else, or driving to other stores to buy different things.

The bigger problem may be that we all want everything, and we want it now. Some preliminary research has shown that the shorter the delivery window—within one or two business days— the greater the environmental impact of the delivery. If I say I want my delivery tomorrow, it may travel on a plane, which produces nearly 30 times more greenhouse gas emissions than if it traveled by rail, and about 5 times more greenhouse gas emissions than if it traveled on a truck.[26] And it also affects the logistics of delivery—with a larger delivery window, UPS or FedEx or the US Postal Service could group my package with other packages coming to my area in the same time frame. Instead, the truck is packed to prioritize my demands (and my convenience) over the efficiency of the system: the truck may go out without a full cargo load or may make a trip just to deliver to me.[27]

On-demand services like Amazon Prime Now, Instacart, and others deliver what you want within hours, often making dedicated trips for the things you've purchased, which adds another complicating factor in terms of traffic. Some of those services specialize in food delivery, and when it comes to groceries, that delivery may require refrigerated trucks, which also use energy for cooling.

Some cities and companies have come up with solutions, or at least ways to lessen the number of trucks on the roads or the time drivers spend making deliveries. One example is delivery lockers. Instead of delivering to each individual address, drivers can drop packages off at centralized locations—Whole Foods, 7-Eleven, and Walgreens are some participating stores—into lockers where the e-shoppers can pick them up. Another solution is delivery at off-hours. In New York, the city's Office of Freight Mobility (part of the Department of Transportation) encourages logistics companies to make deliveries between 7:00 p.m. and 6:00 a.m. Delivering goods at night reduced air pollution by 60 percent (preventing more than 6,000 tons of carbon dioxide emissions per year), saved companies money, and led to fewer fines for trucks.[28]

UPS has also introduced electric tricycles in some cities to pick up and deliver packages from stationary trailers: Pittsburgh, Seattle, and Fort Lauderdale in the US; Hamburg, Offenbach, and Oldenburg in Germany; and a similar model in Toulouse, France. Using these electric trikes reduces emissions and relieves congestion, especially in the often dense and old centers of European cities.[29] (Dense and old makes it sound so mean. Sad but true.)

The rise of e-commerce also has implications for commercial and residential energy use. First, there is the question of consumption: people who shop online also end up shopping in stores, whereas people who primarily shop in stores don't also shop online. So those of us who shop online often do both, possibly adding more car trips and more actual stuff to our personal carbon footprint.

Another study, which looked at how e-commerce affected energy use, found that online shopping led to more energy use in residential and commercial areas: every percent of growth in e-commerce adds 48.6 trillion BTUs (British thermal unit, a way of measuring energy) to our national energy use. Let's put that in context: 48.6 trillion BTUs is about 0.05 percent of our national electricity use, which doesn't sound like that much. But if e-commerce, which is currently worth about $474 billion, grows by 1 percent (or $4.74 billion), we add an additional 0.05 percent to our total energy use. E-commerce has grown by an average of 15 percent each year since 2010, adding about 729 trillion BTUs, or an additional 0.8 percent to our national electricity total. At home, people are using more energy to browse online and using the time they saved by not going to the store to do other, possibly more energy-intensive activities. Some of the additional energy also comes from stores trying to make their physical locations more appealing to customers, which might mean . . . a light show. Or an online-only company opening up a physical store.[30]

When I talked to the co-author Florian Dost about this study, we had a good time egging each other on about consumption and the end of the world.[31] Then I ordered some new pajamas online. But we talked about how improving a system usually means that more energy can be used, maybe in new and different ways. And we weren't saying that less energy should be consumed, just that it's currently difficult to escape an increasingly consumptive system as technology and efficiency improve.

It's also, generally, hard to predict where e-commerce is going to go. It's grown incredibly fast in China, but slower than economists and others predicted in the US and Europe. In the US, e-commerce has only been possible since 1991, when the National Science Foundation allowed commercial businesses to operate online, and the rise in online shopping has also coincided with a global economic recession, when people typically spend less money.

The online marketplace has also shifted where our emissions come from, by reducing the need for a lot of physical things, like CDs, newspapers, and DVDs. E-commerce may have increased the number of things that are shipped in cardboard boxes, but it's also changed the way we consume a lot of our news and entertainment, leading to the dematerialization of parts of the supply chain.

But it's not clear yet how many more things will be decarbonized or what impacts the move online could have, like we saw with video streaming. Another study showed that downloading video games online actually used more energy than buying them online on Blu-ray discs (which included the commercial and transportation energy costs).[32] And more financial transactions, legal agreements, and other intangible commercial activities take place online,[33] which may save energy and resources, too.

It's really frustrating that I can't come up with answers about whether e-commerce is good or bad for the environment, but I do think that the manner of our preoccupation with its environmental effects is slightly off. The problem isn't the cardboard, despite that being the waste we see. The problem isn't truck traffic replacing car traffic, or even supplementing it. And the problem isn't more electricity used at home, because maybe it's better to be using more electricity at home than in stores or factories. The problem is how all of these things work together. We're buying more stuff (which we may not always need), which requires more things to be shipped (maybe in more absolute numbers of boxes, even if the amount of cardboard stays the same), which means more trucks on the road, which means less time spent driving to the store, which means more free time, which means more time driving somewhere else, maybe to buy something else, or more time watching a movie at home, which we stream from the Internet.

The problem isn't e-commerce necessarily; the problem may be us.

Silicon Valley: A Toxic Waste Dump? You Decide

Before Silicon Valley was the idea center of the Internet, it was a group of factory towns, the blinking heart of "clean" manufacturing, the hallmark of the Information Age. I went out to California in late April, when the flowers were starting to come out, the brief season before much of California turns from green to brown to fire, to find out more about how this place—so closely associated in our minds with the nebulous (though not really) space of the Internet—connects to the earth.

Sometimes it feels hard to remember that Silicon Valley is an actual place, a collage of parched suburbs, and not just the collective noun for information technology companies. A gaggle of geese, an exaltation of larks, a Silicon Valley of start-ups. Or something close to that.

Now, though, the area trades mostly in the rarified and intangible realm of apps and software, but it was a major industrial center for much of the twentieth century. Semiconductors and microprocessors rolled out of factories scattered all over the area (known on maps as Santa Clara County) from the 1950s to the early 1990s—AMD, Apple, Applied Materials, Atari, Fairchild, Hewlett-Packard, Intel, National Semiconductor, Varian Associates, and Xerox, to name just a few. From the mid-1960s to the mid-1980s, Santa Clara County added 203,000 manufacturing jobs, with 85 percent of them in high tech.[1] Beginning in the 1980s, as government contracts disappeared, Silicon Valley companies moved toward creating software, and beginning in the 1990s and up until now, companies there have largely focused on Internet-based applications.

It's hard to see that now, when glass-walled office buildings, corporate campuses decorated with primary-colored jungle gyms, and strip malls along the seemingly-hundreds of highways that come together as one and then bloom into concrete clovers, connecting Silicon Valley to every other conceivable part of California, dominate the landscape of this former industrial area.

But all of that industrial history left something behind: toxic waste.

During my reporting, I came across Amanda Hawes and Ted Smith, two occupational health and safety lawyers who have been suing tech companies on behalf of workers sickened by working in tech factories and their children, who often suffered as a result.

They invited me to come visit them in San Jose, where they live, and we picked a weekend in April because they were also going to be hosting a documentary film crew from South Korea also coming to interview them about the toxic legacy of the area. South Korea, now home to some of the world's largest chip manufacturing companies, is dealing with many of the same issues that Hawes and Smith and countless workers dealt with a generation ago.

I was skeptical about how the filmmakers and I would conduct simultaneous interviews with Hawes and Smith and some of the people they had defended over the years, since reporting for writing and video can be pretty different and they were making a documentary in Korean, or how exactly it would work to visit all the sites together, but by that point I had my tickets, so I went.

I met up with Smith and the film crew at a bagel shop in a strip mall in Sunnyvale for our road trip around the former factories, now office buildings/toxic waste sites. It had been decided that I should sit in the middle seat during the road trip for reasons that were not explained but seemed nonnegotiable. My skepticism about the effectiveness of simultaneous book/documentary English/Korean reporting had been confirmed, because no one else seemed especially interested in talking about the environment, but those concerns quickly became overshadowed: within seconds of getting in the car, one member of the team asked me if I was familiar with *Eichmann in Jerusalem* by Hannah Arendt, because wasn't it just so obvious that Silicon Valley was the perfect encapsulation of the concept of "the banality of evil"? Despite usually feeling confident discussing subjects I have absolutely no knowledge of, and having never learned how to pronounce "Arendt," I became nervous and somewhat hesitant to engage. Also, carsick. As a result, I simply nodded. Now that we were all in agreement, we moved on to the subject of government contracting of tech companies, which also turned out to be evidence of both the banality of evil and a manifestation of capitalist fascism. I tried to focus on the sites we were visiting, but I was now preoccupied with larger questions about my identity: I hadn't seen direct links to fascist capitalism, so by the time we got out of the car, I was worried that maybe I am actually, secretly, a pro-capitalist, blindly patriotic titan of industry, working for The Man.

Anyway, we drove around Santa Clara County, and from the middle seat, I tried to understand how this place became one of the most polluted counties in the country, in terms of the number of Superfund sites. It didn't look polluted, and all of the former factory sites were now occupied by their higher-tech replacements—Google, Microsoft, and others—but these places had never been fully cleaned up.[2]

So how did it get to be this way? When you see the Google Quad Campus, you wouldn't think that Google would potentially expose its engineers and sales and ad reps to toxic chemicals. Plus, it looks way too nice to be a Superfund site: There's a pool with primary-colored umbrellas, matching bikes, and a contained universe that looks more like a college or a park than a satellite campus of one of the biggest companies in the world.

But it turns out that this idyllic garden of corporate harmony is also part of the Fairchild Semiconductor Corporation (Mountain View Plant) Superfund site, an EPA designation that denotes some of the most contaminated or polluted land in the country. It was finalized as a Superfund site in 1989 and added to the National Priorities List in 1992. While thousands of tons of contaminants have been removed, it is still being cleaned up.[3] But you wouldn't know it. A friend of mine who worked at Google Quad 1 had no idea that the office was on top of a waste site, despite a 2013 incident where toxic vapors got into the buildings, exposing the office workers there to levels of chemicals above the legal limit set by the EPA.[4]

There are twenty-two other active Superfund sites in Santa Clara County, almost all of them a result of toxic chemicals involved in making computer parts, all designated as such in the mid-to-late 1980s. They remain active Superfund sites because completely cleaning up these chemicals may be impossible.[5] These sites came to the attention of the EPA after groundwater testing in the area revealed that toxic chemicals—notably, a solvent called trichloroethylene—were present, possibly from leaking pipes or underground storage tanks.[6] Trichloroethylene, which was used to clean the semiconductors (a component of computer chips) during the production process (later replaced with a "safe substitute" that also caused problems),[7] is associated with increased risk of certain cancers, developmental disabilities among children exposed in utero, increased rates of miscarriage, and endocrine disruption.[8]

At the time, IBM, Fairchild, and other companies accused of

polluting the groundwater denied that the chemicals posed any sort of threat to human health. In 1985, a California Department of Health Services study reported a significantly higher than expected rate of miscarriage and birth defects near the leaking tanks, though they did not have conclusive evidence to tie the health problems to the leaks.[9]

Over the last three decades, the EPA and the companies involved have filtered and cleaned some of the affected groundwater, but the EPA website acknowledges that cleanup will continue for many decades. That's in part because when the toxins get into the groundwater, they do what water does: flow. Now, the toxic molecules have spread throughout the area—NBC Bay Area reported that, according to government officials, there are 518 toxic plumes of groundwater in Santa Clara County, a number that advocates like Hawes and Smith say is too low.[10] The EPA maintains that there are no direct exposure pathways to the contaminated groundwater anymore, but the problem now is that the toxins can get into the air in buildings via a process called vapor intrusion, which is what happened at Google. It's not entirely clear yet what the result of lifetime environmental exposure is, but occupational health lawyers and activists like Hawes and Smith have spent decades fighting for the people who worked with these chemicals in Silicon Valley's factories.

Hawes and Smith have represented tech industry workers for decades. They began in the late 1970s, when they identified a cancer cluster among workers in Santa Clara County and, with EPA funding, brought attention to the toxic waste in the county. They saw high rates of miscarriage among workers as well—two times the national average—confirmed by two separate industry-funded[11] studies in the 1980s. Hawes has since represented dozens of workers and their children who were harmed by occupational exposure or exposure in the womb. She describes the amount of toxic waste in Silicon Valley (where she lives—in San Jose) as "mindboggling."[12]

Tech companies said that they stopped using these chemicals, but since most chips are now manufactured overseas, though contracted by Apple and others, it's harder to enforce and keep track of which chemicals are and are not used.

Bloomberg Businessweek reported that miscarriage rates among women who worked at some of these factories in South Korea, home to Samsung and SK Hynix, which together controlled 74 percent of the global chip market in 2015, were three times the rate of the general population; two young women who worked next to each other at one factory were diagnosed with the same form of aggressive leukemia and died within months of each other. This form of cancer otherwise affects three out of every 100,000 South Koreans every year. It's hard to say whether these chemicals have truly been eliminated, especially with evidence like that. It's harder to hold smaller manufacturers accountable, or to even know where or who they are. Some factories are also in countries with less stringent regulation of chemical and environmental hazards, like China and Vietnam.[13]

The health and environmental hazards of the tech industry are separate from our lives and from the way that we use these machines. Using computers and cell phones—which require semiconductors or microchips to work—has become so essential to life all over the world that it's easy to ignore the problems with building them. But it's clear that the electronics industry is as much about chemistry—global chip manufacturers buy $20 billion worth of chemicals every year—as it is about computing.[14]

Now we'll look at what these devices are made of—explorations of the periodic table in store!

Mining for Tech

In central Africa, far from the corporate headquarters of Silicon Valley or the factory floors in China and South Korea, lie some of the richest rivers of minerals in the world. In Congo, workers tap the cobalt veins that spread beneath the soil—an astounding source of wealth in one of the poorest countries in the world—hauling up this precious metal and sending it off to be used in lithium ion batteries, the invention that, more than almost any other, has enabled the portable technology revolution and powers our phones, laptops, and electric cars.

The economy of Silicon Valley and the tech universe more broadly is powered by the lithium ion battery. Similar to the dominant narrative about Silicon Valley, the one about lithium ion batteries is that they are clean and efficient. That's partly because, unlike earlier batteries, they don't use lead (toxic) and acid (corrosive), and because they are light and rechargeable and can store more energy, so we tend to overlook or ignore the hazards of producing them. Though much of the demand for the elements and other materials these batteries require is driven by the electric car market, the same principles apply to batteries for electronic devices, just in smaller amounts.

Lithium ion batteries have changed the way we use technology: we can have cell phones and laptops because the batteries can be shrunken down enough to power an iPhone, and they are rechargeable. And like many different parts of different technological devices, they are made of difficult-to-come-by metals and other materials, which often have to be extracted from below the surface of the earth, often at great environmental and human cost. There are a few different types of lithium ion batteries, but the most common ones now use a cobalt cathode—making cobalt the main active component of the battery—requiring large amounts of cobalt, as well as lithium and graphite.

There are estimated to be about 100,000 workers in Congolese cobalt mines, including children, who toil hundreds of feet below the earth's surface, earning up to $2–3 on a good day, despite the high price of cobalt (on average, between $20,000 and $26,000 per ton in 2016). These miners (they call themselves *creuseurs*, or diggers) chip away at the cobalt supply often using hand tools, without protective clothing or equipment or other safety measures. Thirteen miners were killed in a mine collapse in 2015; in 2014, sixteen were killed in landslides and fifteen in an underground fire.[1]

These poorly regulated mines present health hazards to those who work in them and who live near them: heavy metals, like lead, cadmium, and uranium, long buried, make their way up to the surface or into the groundwater and have been shown to cause respiratory problems and birth defects in exposed populations.

Cobalt is the most expensive element in lithium ion batteries, and nearly 60 percent of the global cobalt supply comes from Congo, with most of it processed in China. In 2016, the *Washington Post* traced cobalt from small-scale Congolese mines to a single Chinese company, Congo DongFang International Mining, part of one of the world's biggest cobalt producers, Zhejiang Huayou Cobalt. Zhejiang Huayou Cobalt supplies the cobalt for batteries that end up in all kinds of products, including, at times, iPhones. Apple acknowledged to the *Post* that cobalt sourced from Zhejiang Huayou Cobalt was used in some iPhone batteries, and admitted the possibility that some of its cobalt was mined in these unsafe and illegal conditions.[2] In response to the article, Apple promised to increase scrutiny over where its cobalt comes from and to clean up its supply chain. In 2017, they published a list of their smelters[3] and said they would stop buying cobalt mined by hand from Congo.[4]

While cobalt is not currently classified as a conflict mineral, some say it should be added to the list: it may not be funding war directly, but the potential for human rights abuses in mining and the difficulty in tracking the supply chain suggest a clear need for regulation. On the other hand, as researchers Sarah Katz-Lavigne and Dr. Jana Hönke have argued, regulating cobalt like a conflict mineral risks replacing dangerous income with no income at all for workers in Congo, and could have the unintended effect of pushing cobalt smuggling and operations further into the darkness.[5]

Some lithium comes from the Atacama region in Chile and in the salt flats of Argentina, harvested from lands where Indigenous groups hold surface and water rights, though the vast profits from lithium production are not shared with them. These communities struggle with sewage treatment and drinking water and are often unable to heat their schools.[6]

Lithium mining also requires a lot of water—one plant pumped 2 million gallons in a day—and the Atacama region is among the driest in the world. On average, the area typically sees 2 inches of rain per year, though some places receive as little as .04 inches;[7] the presence of lithium mining operations depletes an already inadequate water supply for Indigenous communities.[8]

The last main ingredient is graphite, the majority of which is both mined and refined in China. Dust from those processes leaves the air sparkling, buries crops, cakes the landscape in soot, and pollutes the water, rendering it undrinkable. According to the same *Washington Post* series, the river in one of these graphite-producing areas no longer freezes because it is so polluted.[9]

We've been talking about only batteries, and already the environmental and human consequences are clear. The demand for computers, cell phones, televisions, and so many of the other electronic devices that we use every day contributes to the impact of mineral and resource extraction. If you were hoping that the waste and pollution ends with the production of the devices, it doesn't. It also extends into how they use energy and how they're thrown away, too.

Vampire Power

There once was a time when it was actually possible to turn an electronic device off. Those days are over. Of all the electronic devices you have in your home—around sixty-five in the average American home—about fifty of them (about three-quarters) are always drawing power, even if you're not using them or think you've turned them off.[1]

These devices—like coffee makers with digital clocks—may be using only a little bit of electricity, in most cases, but the number of devices and the required constant supply can add up. Plus, some devices, like your cable modem or router, are always on, even when you're not home or not actively using the Internet.

In a study of homes in Northern California, it turned out that about a quarter of residential energy was being used by devices that were "off" or in standby or sleep mode, what might be called vampire or idle power. If that held true across the country, that would add up to the output of fifty large power plants and $19 billion in electricity bills every year. Spread out over a lot of people, it's not a lot for each one of us, but the point is the aggregate: as a society, we're just throwing energy away. And that doesn't even count the number of computers and other devices in offices, banks, and stores that stay on or go to sleep when workers head home.[2]

Until 2016, the electricity generation sector represented the biggest source of carbon dioxide emissions—that year, it was replaced for the first time by transportation-related emissions. But that's not because we're using less electricity—it's because it's coming from cleaner sources (less coal, basically).[3] We're using much more electricity than we used forty years ago. In 1978, each American used about 9,560.5 kilowatt-hours of electricity each year.[4] To put that in some context, that amount of energy is enough to keep a sixty-watt equivalent LED light bulb (which uses 8.5 watts) on continuously for about 128 years.

In 2014, the last year that data was available, we each used 12,984 kilowatt-hours.[5] That per capita number has actually stayed relatively constant since the early 2000s, and the cleaning of the grid, through the addition of natural gas and renewable energy, has made a big difference in reducing the amount of emissions. But we're still using more, and there are a lot more of us. In 1978, there were about 222.6 million people in the US. By 2014, there were 318.5 million, resulting in an increase in electricity consumption of about 2 trillion kilowatt-hours.*

Some of what this captures is industrial, commercial, and transportation-related electricity consumption, so that doesn't necessarily have direct bearing on the way that you or I use energy. However, while the world's population has grown, the world makes more stuff (requiring more electricity) than it used to.

Now that I have revolutionized your understanding of electricity consumption, we should probably talk about residential energy consumption, since that's where much of our idle power load is coming from, and is what we're most directly in control over.

---

* And this rapid growth in electricity use isn't just an issue in the US. In China, over the same period, the kilowatt-hours per capita went from 199 to 3,927, while the population grew from 943 million to 1.364 billion, an increase of 5.1 trillion kilowatt-hours. In Nigeria, which has seen its population nearly triple since 1977 to more than 176 million, electricity consumption per capita remains relatively small—144.5 kilowatt-hours per person—but has grown from around 3 billion to 25 billion kilowatt-hours over the same period.

In terms of residential consumption, each American household uses about 10,339 kilowatt-hours each year. If idle load is about a quarter of our total electricity consumption, then that's about 2,584 kilowatt-hours per house. (To put that back in our very useful metric of how many years you could power that light bulb: around thirty-four years.)

Since the light bulb thing is not working for me and likely not for you, either—although who knows, maybe you measure everything in terms of light bulbs—I thought it might be helpful to talk about it in terms of different devices you might use. Also, light bulbs (unless they are smart bulbs with a Wi-Fi or Bluetooth connection—stay tuned for the Internet of things discussion!) don't use idle power, so it's harder to explain how exactly you could be using electricity without knowing it.

Cut to me, lying on the floor of my apartment with my watt measurer in hand, trying to figure out how much electricity my devices use in different states of on, idle, and off. I took some action shots, but I'm not sure they totally capture the thrill of the chase. I will leave you to imagine the excitement, the passion, and the general hysteria that characterized this experiment.

I used a Kill A Watt power meter to measure the electricity being drawn from my devices and some from friends' houses. Here's a tip—most people don't like you going over to their houses and unplugging and replugging and then unplugging and replugging all their devices, but unfortunately for them, I'll stop at nothing to discover the truth.

There were some devices that used just as much power when they were on as when they were off. You may have heard about cable boxes being enormous power hogs—and they are. Mine used 28 watts when it was on and recording a show, and 26 watts when it was off and not recording. If I didn't turn my TV on once this year, I would still consume about 227 kilowatt-hours, which is more than the per capita energy use in several countries, or Kenya and Haiti combined.[6]

I have a first-generation Apple TV because my husband said we didn't need a new one, but the problem with using this relatively old device (about twelve years old) is that while it still works, it uses energy pretty inefficiently: 21 watts when on, and 17 watts when off.

As of 2018, 65 percent of American adults use a high-speed Internet connection at home, which usually entails at least one modem and router.[7] My router and modem, which together provide me with Internet access, are always on. Combined, they use 19 watts, which is not a lot, but in a year, that's another 166 kilowatt-hours, and most of the day, we aren't even using that energy —we're out of the house or asleep. If we took my power, and assumed it was roughly the same as in all American homes with a broadband connection, it adds up to roughly 20 billion kilowatt-hours every year.

Those appliances staying on makes a little bit of sense; if you think about it for a little longer than normal, you'll understand why they have to be on: a cable box, an Apple TV, or a Sonos speaker system needs to be on to receive a signal from the remote control. The question there is, do we really need them to be ready to go at a moment's notice, as soon as we walk in the door or decide we want to hear a song?

And the extreme convenience stretches even further. Now, traditional appliances that didn't have much need for idle power can hop online, too. Some mattresses, light bulbs, refrigerators, and ovens come with a Wi-Fi connection. An app on your phone can turn your microwave on, preheat your oven, warm your mattress, and set some mood lighting before you even open the door.

If you ever hear tech enthusiasts or scientists talking about the Internet of things, this is what they're talking about—a network of

Internet-enabled appliances, in a domestic context, and machines, vehicles, and more in an industrial one. A lot of concerns about the Internet of things come from its implications for privacy: these devices receive signals from you about how you want them to operate, but they also send the information about how you're using them to the company's cloud server, where that information is stored, and perhaps sold and shared.

The problem from an energy perspective is that in order to receive that command from your app, these devices have to maintain a constant Internet connection, which means that they also have to be on. They probably aren't drawing a lot of power, but as more and more devices come online, some scientists warn that the idle power load from all those small connections could be creating an accidental significant demand on our electricity sources. Plus, the growth in data that the Internet of things will also bring has more electricity implications: we will require more and more data storage, requiring more servers and more electricity to power all of our devices. Of course, lots of devices are also much more efficient than they used to be. As battery technology improves, laptops use electricity more efficiently than their predecessors— desktop computers—and cell phones and tablets do also. If used correctly, Internet-connected devices like air-conditioners can be set up to be used more efficiently, offsetting or overcoming the idle power consumption.

There's nothing inherently wrong about harnessing the power of the Internet and digital technology to make our lives "better" or "easier"—if we define those two things by the convenience or save time and effort afforded us by not having to turn on our coffee maker directly. But there's an ease with which we turn our lives over to these devices, and sometimes the focus on what that does to our brains or our privacy overshadows the implications for how we use electricity, which is that we're using more of it, and we're using it almost thoughtlessly. It would be great if this kind of electricity consumption—the tiny, constant kind—figured into our calculations of how we use electronic and digital devices.

If there's a chance that reading this made you concerned about your electricity use, this is one situation where there are a few things you can do. They aren't going to dramatically reduce the overall global electricity consumption, but maybe they will make you a more conscious consumer.

If you use your gaming console to stream video, stop doing that! Those things can use up to forty-five times more power than an Apple TV or other streaming devices, according to the NRDC, largely because they aren't good at adjusting their power use to the task at hand (they might use a similar amount of electricity to stream video or play a massive multiplayer online game, which require very different amounts of electricity).[8] It's not enough electricity to neutralize the rest of your environmental impact, but it is something.

I also realize that implying that you should get a different device to stream videos after I just told you about all the resources required to make any electronic device is confusing. It is confusing! It's really hard to know the right thing to do. Using your electricity-guzzling game console to stream movies is probably, overall, less impactful to the planet than buying a new device altogether. But the important thing is to put any of these devices and all of the resources they use into context. I know that's not a satisfying answer. But it turns out there aren't really very many satisfying answers. Sorry.

Some utility companies will provide hourly data for electricity consumption, so you can see when there's a smaller demand on the power plants in your area. If you use electricity-thirsty appliances at this time—like a clothes dryer, which, by the way, you may want to think twice about using—there's a smaller chance that your dryer (or other people's devices) will be powered by dirtier sources of energy, which kick in when the demand on the system is greater than the cleaner sources are capable of meeting because they can provide reliable and constant power.

In some parts of the country, utilities have installed smart meters, which allow you to track how much electricity your home is consuming.

There are even simpler solutions, which you may already be employing: you can use a power strip to group appliances. That way, you can turn everything off at the same time (and actually off, because turning off a power strip is like unplugging, not just like turning off the switch on the device itself). The problem with that is that you may have to reset things—the clock, the Internet connection—or could lose some data.

Finally, let's look at what happens when you have drained these devices dry, or else have decided that you need to have the new iPhone.

The Tech We Throw Away

Americans want to watch the Super Bowl in the highest definition possible. In 2016, that meant buying 8.6 million new TVs. Some of these televisions are returned after the game is over, but a lot of them aren't. And that means that there are a lot of perfectly good televisions, the ones that were replaced, that are disposed of in the early days of every year.

TVs form a sizable portion of the electronics that we throw away every year, part of a category of garbage known as electronic or e-waste. In the US, the most recent data shows that we threw away 3.14 million tons of electronic devices in 2013;[1] the United Nations estimated that, globally, 41 million tons of e-waste was generated that year, and they expected that number to climb to 50 million tons by 2017.[2]

The waste itself is not necessarily the problem. Although producing all of these devices requires lots of energy and resources—as many as forty elements are used to make a smartphone—it would seem slightly more acceptable to throw them away if we used them until they didn't work anymore. But that's not exactly the case. A lot of the time, it's cheaper to fully replace a device than to get it fixed, and tech companies rely on planned obsolescence to make their companies thrive.*

---

* Planned obsolescence refers to a company like Apple releasing a new device with a software update for all of its devices, but the new software is optimized for the newest version, so your older device runs more slowly and so you decide to get a new one.

But what happens to your iPhone or computer when you throw it away? It depends, mostly, on the consumer's responsibility and knowledge. You can take some devices back to electronic stores, where they will, ideally, get the product to an approved recycler. The US has e-waste recycling laws, but not everywhere has dedicated collection services, which should ensure that e-waste gets to the proper recycling facilities.

But most of the e-waste in the US isn't recycled, and the same is true for the rest of the world. Globally, nearly 70 percent of e-waste is unaccounted for, meaning that it doesn't end up in formal recycling centers.[3] According to data from 2011, in the US, only 20 percent of e-waste is formally recycled, which includes waste that is legally exported.[4] The vast majority of electronic waste around the world is dumped into landfills, burned, or illegally exported to other countries where it is often "informally recycled," or taken apart by hand. Until 2017, in the US, the 20 percent that is formally recycled included the legal export of e-waste to India and China.[5] In 2018, China stopped officially accepting e-waste, and other kinds of waste, from other countries.[6]

Most of the e-waste shipped overseas from Europe and North America heads to developing countries in Southeast Asia and Africa (though a growing percentage is originating in Asia), where hundreds of thousands of people support themselves by taking apart the iPhones someone didn't want anymore.

Some of this happens legally, but that doesn't mean it is done responsibly. Loopholes in current regulations allow for e-waste to be exported to developing countries as "donations," but a lot of the time, what is sent is broken and can't be reused or fixed. The United Nations estimated in 2014 that only 15.5 percent of global e-waste was recycled to the highest standards.[7]

So what happens to all of those electronics, which are composed of valuable, often rare, materials? A mobile phone contains a substantial amount of copper,[8] and other electronics contain not insignificant amounts of other precious metals. Circuit boards contain copper foil, which is taken out once the plastic covering is melted. Gold, silver, and other metals are also on the circuit boards. Those are recovered by dunking circuit boards in open-pit acid baths. The acid baths used to extract plastic from the circuit boards often contain cyanide. Computers, televisions, phones, and all other kinds of electronics are also simply smashed by hammers, chisels, and other tools.[9] Cables are burned to remove the plastic insulation and extract the copper. Burning plastic can also release heavy metals, like lead, cadmium, and mercury.

Almost a quarter of e-waste from around the world goes to the same seven countries where there is no oversight of how the recycling is done or whether people and the environment stay safe and healthy.[10]

At Agbogbloshie, a commercial district in Accra, in Ghana, black smoke hangs thick, the product of burning plastic from e-waste for the better part of most days.[11] In Guiyu, China, the rivers run thick with metal.[12] Vast swathes of the developing world are awash in these chemical by-products of the "recycling" of electronic waste from developed countries in Europe, North America, and Asia, and communities and ecosystems are suffering.

In all of these places, women and children are the most likely to work with e-waste. According to the International Labour Organization, children are prized for their small, dexterous hands, which help them take apart the devices. But children are also more vulnerable to the health impacts of these toxic pollutants: kids take in more air, food, and water in proportion to their size than adults, and they have a decreased ability to detoxify themselves.[13]

In a study published in 2015 in *Annals of Global Health*, the authors, which included several scientists affiliated with the World Health Organization, noted some of the negative health effects of exposure to e-waste. Are you ready? "Changes in lung function, thyroid function, hormone expression, birth weight, birth outcomes, childhood growth rates, mental health, cognitive development, cytotoxicity and genotoxicity." The hazardous chemicals released in e-waste recycling "may have carcinogenic effects and endocrine disrupting properties that could lead to lifelong changes due to neurodevelopment anomalies, abnormal reproductive development, intellectual impairment and attention difficulties."[14] Brominated flame retardants, which are often applied to plastics to make them less flammable, can stay in the environment for a long time and "reportedly lead to impaired learning and memory function; altered thyroid, estrogen and hormone systems; behavioral problems and neurotoxicity."[15] Mercury and cadmium can cause damage to the brain and nervous system, and kidneys and bones, respectively. Studies from China reported increases in miscarriages, stillbirths, and premature births, and reduced birthweight and birth lengths. People in e-waste towns had evidence of greater DNA damage as well.[16]

Sometimes, the by-products of destroying the e-waste or the chemicals used to do it are dumped into rivers. Whatever is left over can also leach toxic chemicals into the soil. Both dumping and leaching can cause these pollutants to enter the drinking water supply or the food chain, eventually collecting in the bodies of those who consume them, in a process known as bioaccumulation. Given the risks and the capacity to properly recycle e-waste and how valuable it is, it seems . . . crazy (in a WTF way and in a profoundly sad way) that governments and tech companies have largely abandoned responsibility for this problem and that millions of vulnerable people are exposed to the damage of our collective tech addiction. The problem still exists that, in many parts of the world, this is a viable way to make a living, or at least eke out an existence.

So why does it happen? First, in the United States, it's cheaper and sometimes even profitable to send containers full of waste across the ocean to countries willing to buy our garbage (formerly China, now not China, since China no longer wants to import our garbage and we are looking for customers) than it is to transport that waste domestically to one of the few sites that can properly recycle it.[17]

Second, there has been a perverse incentive for countries like China to take waste from other countries and, as electronic consumption in China (and therefore waste) grows, to allow a cottage industry of recycling e-waste to prosper. Electronics contain a lot of valuable materials that exist in finite supplies—they're elemental metals and rare earth materials, which require energy and resources to extract from the earth. Close to half of the copper supply in China comes from recycled material; before recycling restrictions took effect there in 2013, the e-waste recycling "industry" in Guiyu yielded 20 tons of gold every year (about 10 percent of the US's annual gold output).[18] In other words, e-waste is important to the Chinese economy.

It would be better for everyone if we had more sophisticated recycling practices. Not only would we be able to avoid the incredibly damaging health consequences, but we might also be able to reuse and refurbish pieces of the electronics we've decided to throw away. Since the Trump administration is trying to cut the budget for the Department of Energy's programs on developing recycling technologies (as of March 2018)[19] and the EPA's recycling-related research divisions, we might not solve this problem for a while.

In the US, we're good at the physical process of recycling— almost 40 percent of aluminum in North America comes from recycled stocks—when we actually try.[20] And there's good reason to do it: it's much less expensive—in cost and energy expenditure— to recycle existing materials than to extract new ones.

Some companies are also trying to help in this effort. Apple will take back your old or unwanted devices, and they will maybe give you some money (in the form of an Apple gift card) for them, and either repair and resell them or reincorporate them into the supply chain, effectively recycling it themselves.[21] Best Buy also will accept any device for recycling, no matter when it was made or if you bought it at Best Buy. They may charge a fee to recycle large devices, like TVs or monitors, and they have some limits on how many devices you can bring in one day. With their recycling partners, they will either repair and resell your device, break it down for parts, or recycle it. As of December 2017, Best Buy recycled 1.5 billion pounds of electronic devices, appliances, cables, wires, etc. They're aiming to recycle 2 billion by 2020.[22]

There are other incentives: about 97 percent of global rare earth ore used to make electronics and other products is located in China.[23] When China's demand for these materials began to increase, the Chinese government set strict limits on how much could be exported, which has caused prices to skyrocket since the quotas were set in 2011.

Nedal Nassar, a section chief at the US Geological Survey, said that we shouldn't necessarily be worried about the depletion of any particular material, since we don't know how much is in the ground in the first place. The bigger problem, he said, is that we've only been using rare earths for a few decades, and the technology for how we use these materials is moving too fast for recycling to keep up. Even if we recycled 100 percent of cell phones or other electronic devices, we would likely still go after a few materials—copper and gold, for instance—but would still be creating a lot of waste.[24] He thinks that corporations are in the best position to tackle this problem: "If you think about it from a mineral resource perspective, it does give you competitive advantage because you don't have to worry about sourcing the materials again."[25]

On a governmental level, it's not all completely hopeless. There are international agreements and regulations about e-waste: the Basel Convention prohibits the export of any kind of hazardous waste (except radioactive waste) from developed countries to developing ones.[26] The treaty went into effect in 1992, but the United States has yet to ratify it. Another agreement—the Bamako Convention, which went into effect in 1998—forbids the import of hazardous waste to most African countries.[27] Despite these agreements, the export of hazardous waste clearly still exists; in 2018, a report from a network of environmental ministries of EU countries found that about three-quarters of used electronic equipment shipped to Nigeria came from Europe, and about a quarter of it didn't work.[28]

Some companies, realizing that Nassar is right and recycling is cheaper than mining, are trying to do something about it. Mitsubishi Materials is investing more than $100 million in precious metal recovery plants to extract the valuable materials from electronics and, possibly, those found in lithium ion batteries in electric cars.[29] Japan, which has long exported vast quantities of its recyclables to China, is feeling bullish: the organizers of the 2020 Olympics are hoping to forge the medals for the Games from the recovered metals from electronic waste.[30] Literally going for the gold!

In the end, the way we throw away our devices is sort of like how we use them—with little regard for their inherent value, the resources required to make and power them, or how our individual actions spread beyond our little corner of the universe. Just like that, gold becomes worthless, electricity becomes an afterthought, exposure to toxic chemicals and dangerous working conditions become justified. The Internet and our technological devices aren't inherently problematic—it's how we use them that's the problem.

Food

Everybody eats. And everything we eat comes from the earth, in some way or another. Of all the things I'm writing about in this book, the environmental impact associated with food production is maybe the most intuitive: it grows in the environment, requiring things from the environment, and leaving other things behind.

But the thing about global food production (and even food production in the United States) is that it's hard to understand the scale, even if you know that more than 7 billion people have to eat, and pretty much everything that they eat has to come from the earth. The scale of farming and raising animals and fishing and food production is truly massive: about 29 percent of the earth's surface is land and continents, and around 40 percent of the ice-free surface is used for agriculture.[1] Of that, nearly half is used on not feeding people directly, or at all—45 percent of crop calories are used for animal feed and fuel.[2] Emissions from food production range from about 20 to 30 percent of global greenhouse gas emissions, with about 15 percent of the global total coming from livestock.[3] (Of that, about 39 percent is in the form of methane from "enteric fermentation," aka cow burps, though you probably have heard it's cow farts.[4] It's not farts; it's burps. It's 95 percent burps and 5 percent farts. This is important because people always get it wrong. It deserves all the footnotes it's getting.)[5]

Understanding how the food system works and, more importantly, how it doesn't work is really important to understanding what is going on in the environment: rural drinking water pollution, dead zones at the mouth of almost every major river, landfill emissions, transportation emissions, loss of biodiversity, deforestation, overfishing, and so much more.

I knew that I wanted to write about food because it's something that has really interested me since I started writing about climate change. As the person in my house who does most of the grocery shopping and cooking (you're welcome, George), it felt like my impact in this area was something that I was in direct control of, or if not in control exactly, then at least in close contact with. Yet, even as I learned more about this issue—trying to figure out different strategies for determining the carbon footprint of certain foods, writing about eutrophication from excess fertilizer, pondering the universe as I ate more yogurt than almost anyone else on earth—I still felt like navigating the grocery store was really hard if the goal was to feed myself and limit my environmental impact.

I felt like I had heard the same things over and over again about food without thinking about what they meant: red meat is bad for climate change, other meat is maybe less bad, we waste too much food, organic food is good, eating local is good, there are no fish left in the ocean. I wanted to know which of those things were true and if some of them were more important than others. But I also wanted to write about some of these issues in a different way. Lots of us may have heard about red meat, but apart from cow burps, why is it so bad? It's because of the way we grow the feed, which, for the last few decades, has been mainly corn. So I'm talking about beef and other livestock production, but I'm talking about it by talking about corn and not by talking about cows, pigs, or chickens.

Then I wanted to know, how did food waste get to be such an enormous problem? How did we end up in a country where it makes sense, in terms of the market or in the way we grow food, to throw away about one-third of what we produce? It's the way we label food, the way we want food to look, and the way we grow food, too. As the world's population shoots up to 9 billion people in the next thirty or so years, we'll need to increase food production by about 70 percent (to feed a more urban and richer population, which will demand more food and particularly more meat);[6] it seems like wasting food is stupider and stupider.

Celebrities love organic food, therefore I love organic food. No, that's not why, but should it be? Is that why some people care about organic food? Probably. What does it actually mean for something to be organic? Do the celebrities know? Should I tell them? Will it make them like me? Is it better for the environment, and if so, in what ways? Does the organic-industrial complex exclude some farmers to their detriment and our own? (Yes.) Why do we effectively have two farm systems: one that works with the earth but struggles with production, and another that seems to work against the earth but produces an enormous amount of food?

I had heard a lot too about "eating locally" when I was in high school—that it was a meaningful way to fight what we then called "global warming"—but I stopped hearing about it so much. Why?

Had it been shown to not be that important? If buying locally doesn't necessarily save emissions, does that mean that importing food from all over the world is good for the environment? Or, asked another way, even if it's not bad for the environment, doesn't it seem wasteful to have our food circumnavigate the globe, or at least change hemispheres before we eat it? (Yes.) How did we come to take that for granted? (Because, among other things, we suck.)

What about overfishing? That also seemed like something I had heard a lot about at one point but had pretty much stopped hearing about. Was that because I wasn't listening, or because the oceans are a lost cause, or because things had gotten better? (It may actually be a mix of all three, which has made figuring out how to write about fish really complicated and difficult.) Are there fish that we shouldn't be eating, no matter where they come from? (Bluefin tuna.) How does fish farming fit into all of this? Is fish farming as gross as I think it is, with tanks and nets full of aggressive, fatty fish, flailing and squirming and slimy? (Yes.) Do we need to farm fish? (Yes.)

There is no way that I could write about everything about the food system. It would be twelve books and you would never read them and I would die before finishing the project. I would be the Robert Caro of American agriculture, except I would be dead. I picked these five things to write about because they are complicated and important, and I think they give a sense of the size and scope of the system. There are lots of other things I could have written about. I wanted to write about plastic food packaging, and whether we should actually ban plastic bags. (Plastic bags are a major source of plastic pollution, and if you live in a place with bad waste management where most plastic ends up in the ocean, then you shouldn't use a plastic bag. On the other hand, plastic bags are less energy-intensive to produce and transport than paper bags or reusable bags. In general, we should use less of everything. Similar principles hold true for reusable versus disposable coffee cups.)

So anyway, these are the things I wrote about. They're interesting and complicated and while I tried to present the issues in the most complete way without ignoring their complexities, I'm sure there are some things I missed. Feel free to jot them down in the margins and then go ahead and keep it to yourself.

The Greediest Crop

If you had asked me before I started writing this book or even when I started reporting this section of the book, "What is the biggest environmental problem created by agriculture?" I would have said that it was how much red meat Americans eat or pesticides or something. They are problems, but they're not it. It's corn. Other things, too, but mainly corn. We grow so much corn that climate change in the Corn Belt, which covers most of the Midwest, is happening at a different rate and with different effects than everywhere else on earth, in defiance of all the best and most precise models that scientists have come up with. Climate models show that the Corn Belt should have warmed a little and gotten a little wetter. That is not what happened at all: summers there have gotten as much as 1°C colder, and rainfall increased by as much as 35 percent between 1982 and 2011. There's so much photosynthesis (from all the corn and soybeans), which causes evaporation from plants, happening there that it's adding more moisture to the air.[1] It might be the second-best place to grow food on earth, at least for a few months every year, when the Corn Belt is the most productive region on the planet. So why does it predominantly grow corn, something that you probably barely eat?

You eat so much of it you don't even know. In fact, Americans eat so much corn—in the form of meat (the food our food eats), syrup, oils, and alcohol—that according to scientists, on a molecular level, Americans are basically "corn chips with legs."[2]

But there's also a lot of it we don't eat, or at least not directly: just under forty percent is animal feed, and another thirty percent is used for ethanol. Of all that corn, we only eat about 1.5 percent of it in the form of actual corn (or popcorn or cereal.)[3]

I realize I am not the first person to suggest that we unconsciously eat a tremendous amount of corn, but I promise you that I am not (entirely) ripping off Michael Pollan, whose book *The Omnivore's Dilemma* was hugely important to me in writing this section, and whose other book *The Botany of Desire* changed how I think about plants and people.

Cornfields take up 10 percent of all our agricultural land (about 90 million acres), an area almost as big as California.[4] Since 1919, the annual corn yield has gotten about seven times bigger, though the number of acres of corn planted is pretty much the same.[5] We grow so much corn partly because corn is amazing. Native peoples in Mexico cultivated corn between 5,000 and 8,000 years ago, Indigenous groups throughout the Americas planted it for generations, and without corn (and more importantly, without the Wampanoags, who taught them how to grow and harvest it), English settlers in seventeenth century Massachusetts would not have survived. Corn can grow almost anywhere. It's more resistant to hot and dry conditions than other plants. It has greater yields than other grains and can be converted into more products: alcohol, oil, animal feed, human feed, flour, sugar, and more.[6] Chemicals derived from corn are used in so many things: detergents, tobacco, aspirin, paper, surgical sponges, and dynamite.[7] Corn was foundational to the pattern of settlement and early markets of the Midwest, which in turn shaped how the economy worked and how the land looks.[8] Now, the land looks like corn. In British English, *corn* technically means "the chief cereal crop of a district," so it doesn't always mean "maize." The way the US has grown corn has literally changed the meaning of the word for Americans.[9]

And because of our national ethanol mandate, known as the Renewable Fuel Standard, which requires a certain percentage of the gasoline that goes in our car to be made of ethanol (oil from corn), there is always a market for corn.

So what does it mean for the environment that so much of our land is taken up by a few crops—and taken up by those crops every year, year after year? And not only that, but what happens when the land has been turned into a giant factory using millions of tons of fertilizer, lakes' worth of water, and more than 1 billion acres of land producing enough food to throw away half of it? (The food that we actually eat only takes up about 77 million acres of land.)[10]

In a phrase: outlook not good. In an extended explanation, it means a few things: a loss of biodiversity and native ecosystems; poor soil health, which makes it harder to grow everything; more fertilizer (because of that poor soil health), which pollutes groundwater, drinking water, oceans, rivers, and lakes. This is going to be pretty broad strokes, but it will get us set up for the rest of the section.

Biodiversity and Native Ecosystem Decline

Iowa, which produces the most corn of any state, has lost 99.9 percent of its pre-European settlement tall grasses. The rest of the western Corn Belt (North Dakota, South Dakota, Nebraska, Minnesota) has lost about 99 percent, and that region is losing grass cover (and gaining more corn and soybeans, primarily) at an astonishing rate: according to one study, from 2006 to 2011, grasslands in the Corn Belt (including grazing lands) were converted to cropland at a rate comparable to the deforestation of Brazil, Malaysia, and Indonesia. Those are countries where major expanses of forest are burned and logged to make way for agriculture and met with lots of attention from Greenpeace and other environmental organizations.[11] In the Corn Belt, it's not.

The land that's being converted to cropland is often not land that should be cropland: it might be vulnerable to flooding and drought (Yeah! Both!) or wetlands, which are important habitats for prairie waterfowl.[12] Getting rid of grasslands and wetlands also means more than an important loss of habitat and natural ecosystems: another result is greenhouse gas emissions. The carbon stored by these plants and in soil gets released into the atmosphere, and they're no longer there to store future carbon. During the harvest, crops get cut down every year, releasing more carbon dioxide. One study found that clearing land for crops (in this study, specifically for biofuels) releases 17 to 420 times more carbon dioxide than what would be saved by burning biofuels instead of fossil fuels.[13]

Planting that much corn and planting it as a monoculture (the practice of growing genetically similar plants over large areas, year after year) also takes a toll on biodiversity. Here are some arguments in favor of monoculture: Twelve plant species make up 75 percent of our total food supply,[14] so we need a lot of them (though monoculture is not necessarily the best way to improve yield). If a farm just grows one thing, growers don't need as much human labor to plant and harvest it, especially if they are using chemical weed killers and pesticides,[15] and there's a more uniform plant structure, since the crops are often genetically identical, which consumers like (I scream, you scream, we all scream for uniform plant structures!).[16]

Here are the arguments against. Monoculture crops are more vulnerable. Extreme weather poses a greater risk to genetically identical fields of crops, since what is dangerous to one is dangerous to all. They're also more vulnerable to disease and insects for the same reason, which can cause total crop failures instead of more modest crop losses.[17] Climate change will make losses due to weather, disease, and insects more likely and more frequent. Agricultural crops rely on hundreds of thousands of other organisms—insects, birds, microbes—for pollination, pest control, nutrient recycling, and decomposing waste. By reducing the number of different species we plant (or by planting in monocultural, factory-like abundance), we reduce the number of organisms those species can support.[18] We take away the habitats of lots of other animals that need those habitats to live, too. Without biodiversity in an ecosystem, species can be lost, creating the possibility for ecosystem collapse.[19] It also leaves us increasingly dependent on fertilizers and pesticides to protect our crops against the vulnerabilities we've multiplied in the system.[20] In the future, farmers may increasingly come to rely on seed banks as the twin forces of industrialized agriculture and climate change wipe out crops or entire species.[21]

Declining Soil Health

Planting corn, a plant with relatively weak roots, and only planting corn creates two major problems for soil. First, it makes the soil less stable, which generally happens with agriculture, since the older, stronger root systems are torn up, but more so with monoculture.[22] The soil is plowed every season, breaking it apart and making it more vulnerable to erosion, which is a big problem in corn country. Second, the soil is not given time to recover, through crop rotation with nutrient-restoring legumes or less-demanding crops like hay, so it loses the life-giving nutrients it should have in much greater abundance.[23]

When the soil has fewer nutrients, plants (which need the nutrients) don't grow as well, so growers put more fertilizer on the soil, which adds nitrogen and phosphorus to the soil, two of the nutrients plants need in order to grow. Fertilizer is usually applied in amounts that are greater than what the plants need—it's been estimated that, around the world, crops absorb only one-third to one-half of the nitrogen applied to soil as fertilizer[24]—so there's lots of it in the soil that isn't taken up by the plants. (Quick note: fertilizer has also helped feed the world, enabling the global population to shoot up from 1.6 billion in 1900 to more than 7 billion today and preventing most of them from going hungry.[25] Producing it is also responsible for 1–2 percent of the world's greenhouse gas emissions.)[26]

When the heavy rains come in the spring (and, as we know, rain has increased in the region by as much as 35 percent over the last forty years, and climate change causes more precipitation generally, and more extreme precipitation events as well), the less stable soil is more vulnerable to flooding, so lots of the soil washes away.

And that connects to the third problem: water quality.

Declining Water Quality

One of the most serious consequences of intensive agriculture is what these fertilizers do to our water supply. Across the country, growers apply 5.6 million tons of nitrogen (in the form of fertilizer, plus another million tons of manure) to the soil; much of that fertilizer is absorbed by the land but a lot ends up in the water[27]—both in the groundwater and in other waterways. One of the ways that happens is that, when the rains come, they wash away the unstable soil, which carries with it all of that nitrogen and phosphorus. As the soil becomes shakier and the rains increase, runoff from cornfields increases too, bringing downstream the unabsorbed nitrogen and phosphorus. In the Corn Belt, the nutrients are carried by rivers to the Gulf of Mexico and the Great Lakes.

When the nutrients get to the mouths of these rivers, they

can cause dense growths of phytoplankton, called algal blooms (nitrogen primarily causes these in saltwater; phosphorus in freshwater). These algal blooms block light at the water's surface, preventing it from getting to plants deeper in the water, which need it to live, and making it difficult for animal predators to see their prey, causing die-offs of both. When the algae eventually die, they decompose, using up the oxygen in the water. This creates an oxygen-deficient (hypoxic) zone, which can no longer support life. When people talk about the dead zone in the Gulf of Mexico, perhaps the most notorious one, this is what they're talking about. In 2017, scientists recorded the largest-measured dead zone in the Gulf of Mexico since they started measuring in 1985—nearly 9,000 square miles, bigger than New Jersey.[28]

How did we get all the way here, without anyone addressing this problem? Well, in 2001, a government committee tasked with solving the dead zone crisis (which torpedoes the regional fishing and tourism industries) set a goal for reducing the size of the dead zone to 1,900 square miles by 2015. When it became clear that the deadline would not be met (the dead zone in 2015 was 6,500 square miles) they pushed back the deadline to 2035.[29] Getting the dead zone to shrink down to that size will require reducing the amount of nitrogen that gets into the Gulf by 60 percent.[30] So far, the Gulf of Mexico Hypoxia Task Force has relied on a voluntary system, in which farmers are asked to try to reduce the amount of fertilizer they use or to rotate in some stabilizing, nutrient-generating crops, but no mandatory limits have been set so far. (Agricultural runoff can't be regulated like other pollutants because it's not classified as a "single source pollutant" by the EPA, since the EPA under the Trump administration suspended the Waters of the United States rule, a move that severely limited which bodies of water are regulated under the Clean Water Act, but I digress...)

However, some studies have shown that reducing the amount

of fertilizer—even by such large percentages—might not be enough. The heavier rain brought by climate change could increase nutrient pollution by as much as 20 percent, even if fertilizer use declines.[31] Plus, other studies have shown that all of the nitrogen and phosphorus that we've been putting in the Gulf since 1950 will keep being a problem, even if no more nitrogen runs into the water.[32]

If the nutrients don't get taken up by the plants or get washed away, they can leach into the groundwater. When nitrogen gets in the groundwater, it produces nitrates, which can cause blue baby syndrome in infants (when their blood can't carry enough oxygen and they turn a grayish-blue). The National Cancer Institute found that nitrates in drinking water were associated with thyroid cancer in women;[33] other studies have shown increased risk of bladder[34] and ovarian[35] cancers. (Fertilizer is not the only source of nitrates, which can also come from septic system leaks and manure.) Water treatment plants can filter out these farm pollutants through some costly processes—reverse osmosis, ion exchange, or distillation. A report from the Environmental Working Group, which collected tests for contaminants from 50,000 utilities nationwide, found that 97 percent of public drinking water systems with nitrate levels at or above five parts per million were in rural areas, serving 25,000 people or fewer, and concentrated in some of the country's biggest agricultural states.[36] (Other microbes and pathogens can be treated by adding chlorine, but that can create another chemical by-product, trihalomethanes, which have been described by the EPA as a "probable human carcinogen" and have produced reproductive harm to mice in concentrated doses.)

Corn also just uses a lot of water, leaving less for everything else. Corn alone uses more water for irrigation, pulled from underground aquifers and rivers, each year than is in the Great Salt Lake.[37] In the drought-prone places where corn is increasingly grown, like Kansas and Nebraska, that's a lot of water. (Climate change causes increases in temperature, and at warmer temperatures, air can hold more moisture, so it sucks more moisture out of the soil, causing it to dry out and creating drought conditions. As a result, when it rains, it can rain much harder since there is now more moisture in the air, which increases the likelihood of flooding. In a flood, the soil can't absorb all of the water, so much of it washes away. That's how you can have drought and flooding at the same time.) And what is left, as we have seen, can be polluted.

Why

If monocultures in general and corn specifically create all of these environmental problems, why do we plant it in such large amounts? Three reasons: government subsidies, ethanol mandates, and meat consumption.

None of this had to be so bad. As with so much of our economy and the way incentives play out, the scale has created the problem and then made it worse. Now, the corn industry is too big (too many growers, too much land) with too many dependents (the meat industry, the fertilizer industry, the biofuel industry) to come back to a size that's reasonable (for our actual needs, for environmental responsibility) without a lot of pain.

None of those industries pay for the ultimate cost of that waste and pollution. The industrial farms or the beef producers don't pay for the effects of the fertilizer on the Gulf's declining shrimp fisheries. They don't buy the bottled water for the rural communities whose drinking water is making them sick. Presidential candidates who campaign in Iowa (site of the caucuses, the first contests in the presidential election) and support the ethanol mandate, which provides a noncompetitive market for corn, aren't concerning themselves with the destruction of local butterfly habitats or the drinking water in Toledo, Ohio, undrinkable when heavy spring rains cause annual toxic algal blooms in the west end of Lake Erie.

As taxpayers, we do. We will pay for all of it: the methane from cow burps; the emissions from fertilizer production; the lack of biodiversity in our fields, oceans, lakes, and rivers; our polluted drinking water. But right now, no one is paying to clean it up or prevent it from happening in the first place, which is infuriating and sad, and such a big problem that it makes us feel powerless.

Not to mention that we spend all this energy, use up all these resources, and create health problems for ourselves, and one-third of the food we produce goes to waste.

Wasting Away

Most Americans don't think that they waste much food, though we are aware that food waste is an issue, generally speaking. I also think both those things. It turns out we are all half wrong. According to the USDA, together, we waste about 133 billion pounds of food each year,[1] though environmental organizations think that number could be higher—closer to 160 billion pounds each year, or an average of more than four hundred pounds per person.[2] Either way, it's somewhere around 30–40 percent of all the food we produce. It comes at a huge financial cost, causes a huge amount of greenhouse gas emissions, and is a huge waste of resources (energy, food, water, fertilizer), despite the fact that about 12 percent of Americans don't have enough to eat.[3]

So how did we get here? How did we start thinking it was okay to buy food largely just to throw it away? I asked Liz Goodwin, senior director of food loss and waste for the World Resources Institute: as the CEO of Waste and Resources Action Programme (a UK-based charity) for almost a decade, she helped the UK cut its food waste by 21 percent over four years.[4] Together, we puzzled over how this got to be a problem. In the postwar period, she said, especially in the UK, food was scarce ("My grandfather hadn't seen a banana in years"). But something happened between then and now. Large-scale migration to cities meant that most people were disconnected from how their food was produced and how much work producing it can be. The relative cost of food is low in the developed world, where it is a small proportion of our budget, and we can get anything we want in a store at almost any time of day, so we don't appreciate its value. "We've lost our ability to cook despite all the chef's programs and cooking programs on television," Goodwin said, adding, "Plus, we've become very busy; our lives and our families don't live or eat in the same way." Some of that might explain how our attitudes changed, but it's still puzzling. Or, as Goodwin said, "Nobody can actually say that food waste is a good thing, but you can get some people being cantankerous and saying, 'It's my right to throw away food,'" which is a stupid argument. "It's become a social norm to waste food; no one thinks it's abnormal to eat half a burger and throw away the rest."[5]

There are a few reasons why this problem persists, including and outside of our dumb behavior.

One of the biggest reasons is food labels. A casual, unscientific survey of myself revealed that all Americans think that expiration date labels are probably in some way regulated by the government or at least verified by the Food and Drug Administration (this would seem to be in their wheelhouse) or the US Department of Agriculture. This casual survey revealed that I was wrong. Not only are date labels on food the Wild West of food safety, but people have been trying to impose some law and order in the old-timey regulatory saloon for forty years.

Beginning in the 1970s, Americans started wanting to know if their food was actually fresh or not, and so some food companies started to put open dates (like "use by" or "sell by"), and the federal government started looking into it.[6]

The Congressional Office of Technology Assessment (which no longer exists) found "there is little to no benefit derived from open dating in terms of improved microbiological safety."[7] As a result, the General Accounting Office (now the Government Accountability Office) started pushing for legislation to create a uniform dating system across the country. Without a national system, different food labels would create "confusion because as open dating is used on more products, it would continue letting each manufacturer, retailer or state choose its own dating system."[8]

So a law was passed and the problem was solved once and for all. End of chapter.

As if, dear reader! No food date labeling law was passed, and the responsibility for labeling was split between the USDA and the FDA, but mostly they don't enforce anything. (The one major exception is infant formula, whose date labels are regulated by the federal government because their nutritional benefits decline over time.) The USDA has limited the wording for labels on meat, poultry, and certain egg products to specific phrases—packing date, sell by, or use before—but none of those terms has ever been legally defined or standardized by regulation.[9]

And more than thirty years later, there are still no national laws about dating foods. (If people can start dating foods, what's next? Will they date their computers?! [Pause for laughs.]) Every state can pass its own laws or not, and many have, but they mostly don't make any sense. Let's take a quick survey: New York has no date label regulations; Florida requires that all milk and milk products "shall be legibly labeled with their shelf-life date," which, conveniently, is never defined in the statute, but apparently is around 10 days after the milk is processed;[10] in New Hampshire, a sell-by date is required for milk but not for cream.[11]

Why does it all matter? More than half of American adults think that food date labels indicate something about microbiological safety (as previously established, they don't), and more than one-third think that the federal government regulates these labels.[12] Large numbers of people (about 40 percent of Americans) say they throw food away frequently because of the date on the container, rather than by what their senses tell them.[13] ReFED, an anti-food-waste group, estimates that standardizing date labels would prevent nearly 400,000 tons of food per year from going to landfills.[14]

But it's not just food labels. According to the USDA, American farmers and growers produce about 215 million tons of food for human consumption each year.[15] They estimate that about 31 percent of that—66.5 million tons—goes uneaten.[16] (Other groups, mostly advocacy organizations, estimate that the number is higher—closer to 40 percent.)[17] Aside from making us feel guilty and wasteful (although 76 percent of Americans say they throw away less food than average;[18] yeah, okay), all of that food waste has important implications for energy and resource use, as well as greenhouse gas emissions. Let's add it all up. Wasted food accounts for 21 percent of freshwater use, 18 percent of cropland, 18 percent of all farming fertilizer, and 21 percent of all the waste going to landfills, which releases methane, a powerful greenhouse gas, as it decomposes. Nearly 3 percent of all our greenhouse gas emissions come from the production or decomposition of food we never even eat. That's about as much as the airline industry, and we don't even get to go anywhere. Meanwhile, one-eighth of the American population doesn't have enough to eat, and less than one-third of the food we waste would be enough to feed all of those people.[19]

We're throwing money away: farmers lose out on earnings from food they can't sell, and the average American household spends $1,800 a year on food they throw away.[20] As a percentage of income, food costs have declined for all but the poorest fifth of society, so they pay more for food waste than everyone else. In the US, we spend less cash relative to income on food than any other country, and we spend more money on food prepared outside the home (restaurants, prepared food at grocery stores, schools, and work), which leads us to think of food as cheap, abundant, and convenient.[21] The negative effects (pollution, emissions, etc.) of food production aren't reflected in the cost of food, but we are the ones who end up paying for them, in the form of health effects and climate change and pollution.

We've actually been having this conversation, but apparently not listening, for forty years. In 1977, the General Accounting Office made a report to Congress. It said, "The US can no longer be lulled by past agricultural surpluses and must consider a future that may contain a world shortage of food. In an environment of plenty, the US has not historically been concerned with food losses . . . plentiful food and low prices did not justify the economic expenditure necessary to reduce loss." That loss, they found, represented "a large misallocation of resources. For 1974, 66 million acres of land, 9 million tons of fertilizer, 461 million equivalent barrels of oil." And that was in the mid-seventies, when 23 percent of food was wasted.[22] Compared to then, we now waste 50 percent more food.

Obviously, it's not just at home that we're wasting food. Restaurants, responsible for 18 percent of wasted food, serve portions that are too big or have menus with too many things on them, so their stocks are full of foods that customers may never order.[23]

And there's a lot of food that never makes it out of the fields in the first place. That's not to place the blame on farmers, but on the food system that forces them to use the growing practices they do.

About 10 million tons of food goes unharvested every year. There are a few reasons why, some of which are more relevant to us (monoculture practices, consumer preferences) than others (labor shortages, order changes, food safety threats)[24] starting with consumer preferences. We may not think that we have exceptionally high standards for food, but we do. We want our strawberries, each and every one of them, to look like the Platonic ideal of a strawberry—that is to say, the color of Snow White's lips, the shape of a heart on the end of Cupid's arrow, with a ring of tender green leaves, dotted with little seeds that don't get stuck in your teeth—and not a lumpy blob crowded with large green seeds that look like small insects. We want that perfect strawberry, and so all of the others get left in the field, since there is no market in this country for extraterrestrial fruits and vegetables. Some fruits and vegetables are also deemed too small or too big. The point is that our aesthetic standards mean that farmers can't sell some of their produce to retailers, which adds up to millions of pounds of uneaten, perfectly good produce each year. A study in Minnesota found that up to 20 percent of fruits and vegetables are considered too large, too small, or not pretty enough to meet conventional commercial standards.[25] Imperfect produce can be used to make processed foods—apples for applesauce, tomatoes for ketchup or tomato sauce, other sauces, and so on—but not all of them.

As we covered in our section on monocultures that everyone will force their children to memorize because of the beauty of the prose and the fundamental wisdom of the insights, crops that are planted as monocultures are more susceptible to extreme weather events and pests. As a result, farmers often have to plant more and more of these crops to hedge their bets. Those quantities are often more than the market can support, and so farmers can be left with mountains of crops that no one can do anything with—it would cost the farmers more to harvest them than they would make from selling them. (This applies less to crops that can be turned into biofuels, because there is always a market for those.) What's ironic and upsetting about that is that we have essentially built a vulnerability into the system—monocultural practices ensure that the crops are more vulnerable to environmental factors, so farmers plant more of that monoculture to offset the risk, all but guaranteeing more of the harmful effects of this kind of industrialized agriculture.[26] (This mainly applies to large-scale industrial farms that are more responsible for monoculture planting; smaller scale or family farms may only grow one crop, but are typically less responsible for this kind of overproduction and ensuing vulnerability.)

Overall, according to the USDA, 4 percent of fruits and vegetables go unharvested every year, though that varies by region, season, crop, etc.[27] The USDA found that 1 percent of broccoli was unpicked from 2012 to 2014, but a smaller survey in California revealed that anywhere from 5 to 20 percent of farmers' broccoli fields went unpicked in a given season.[28] Labor shortages can also account for a significant portion of unharvested produce (not enough people to pick the food), as can recalls related to food-borne illness or threats of illness.[29]

Fruits and vegetables account for 35 percent of the wasted food by consumers and grocery stores, and there are some companies that are trying to address that.[30] Imperfect Produce, for instance, is trying to change our ideas about food beauty: it's also what's on the inside. Based in San Francisco, that hotbed of farm-to-table anarchy (though they also serve Los Angeles, Portland, and Seattle, aka the West Coast liberal media elite, and also Chicago and Indianapolis), Imperfect Produce directly sources fruits and vegetables that are too small, too big, or too ugly to sell from farmers and sells it to their customers (on a subscription model) for 30–50 percent less than grocery stores do. According to Reilly Brock, Imperfect Produce's content manager, some of the items that end up in their customers' baskets are "pretty wacky looking." The company often hears things like, " 'Oh, I didn't know a beet could get that big, or an avocado could be so small or that a lemon could look like a Muppet,' " Brock said. I also didn't know that a lemon could look like that. Most of their produce comes from California—in part, because 80 percent of specialty produce is grown there—but they are trying to source locally.[31] However, customers now want what they want when they want it (a topic we will address later), so not everything can be locally sourced, and some of the no-longer-wasted produce ends up being sent across the country. They estimate that they have saved 30 million pounds of food, 900 million gallons of water, and 91 million pounds of carbon dioxide.[32] Some criticism has been leveled at Imperfect Produce and other companies for reselling food that might not have been wasted in the first place and for sourcing from large food companies, like Dole. The founder of Imperfect Produce said that they are transparent about the ultimate destiny of the food they source—5–10 percent may not have been wasted—and that they do buy from large scale producers. Another criticism is that the company is commodifying waste and making money from food that could be sold to restaurants or canned and processed food companies or donated. The risk here is that growers could be incentivized to "overproduce" and sell the food to companies like Imperfect Produce and others, effectively creating a secondary market.[33]

But it might be more effective to cut down on the amount of meat and dairy that we waste instead. On the production side, according to the Food and Agriculture Organization of the United Nations, about 3.5 percent of meat and less than 2 percent of dairy is wasted, in part because the meat industry is good at using all parts of the animal.[34] Overall, however, meat, poultry, and fish account for a substantial proportion of consumer and retail waste by weight (11.5 percent according to the USDA).[35] Of these, reducing the amount of wasted meat would represent the biggest ecological benefit, since meat has the biggest environmental footprint by far. Dairy is also problematic, and we throw away a lot more of it (19 percent of post-harvest waste by weight),[36] so if we reduced our dairy waste, that would have enormous ramifications as well.

In 2015, the EPA and the USDA set the country's first food waste goal: to reduce the amount of food thrown away by 50 percent by 2040 (a similar goal to the one outlined in the UN's Sustainable Development Goals).[37] Congress improved food donation tax incentives in 2015 and extended them to all types of businesses. However, the Food Recovery Act (the first ever food waste bill), Food Date Labeling Act, and Food Waste Accountability Act all stalled in Congress.[38]

Something needs to happen: not only are the negative environmental impacts of food waste enormous, but the UN estimates that the global population is expected to reach 9.3 billion people by 2040, requiring a 70 percent increase in food production. We already have about 30–40 percent just lying around. We could waste less—planning our meals better so we only buy what we need and then actually eat it would be one way—but experts argue that it would be much more effective to stop waste at the source: by growing less food and not wasting it (not so that there isn't enough for everyone, just to be clear).

Now we'll look at how growing organic food fits into that whole "growing more food" thing.

Organic Food: How Good Is It?

I love apples. As with strawberries, I have eaten apples until I got a rash. I'm not allergic—I just ate that many apples. And yet, I cannot be stopped.

When I find a good apple, I will do anything to get it. Enter Rogers Orchards in Southington, Connecticut. I discovered Rogers Orchards during my senior year of college, when I would travel to Farmington, Connecticut, to do research as a visiting scholar at a library there. I thought it was very cool and I talk about it too often. Anyway, I sought out an orchard where I could buy apples as well as some apple cider donuts, because I'm only human. I bought some Macouns, uncontestably the best kind of apple. Now, every time I go to Connecticut (which is often, because my husband's family lives there), I try to figure out if I can go to Rogers Orchards without driving more than one hour out of my way. I will drive an hour out of my way for these apples, but not more. They are the best apples I have ever had. I have also ordered some online at great personal expense. I maintain that the cost was worth it, but others (my husband) disagree.

But luckily, I did order them, because the second time I did, I received a note with my apples from Greg Parzych and Peter Rogers, who currently own and operate Rogers Orchards. In their note, they said that they are an eighth-generation family farm, "committed to the most progressive and sustainable practices" to prepare their farm for the next generation. They acknowledged that their farm isn't organic, but referred to their apples as "Eco Apples." (To be fair, they knew they were writing to an environmental journalist.)

That made me curious about organics in general, and why people who are clearly interested in and committed to environmental stewardship might not practice organic management, since we generally hear that organic food is good for the environment. So I thought I would find out what it actually means to be organic, and if organic food is, in fact, better for the environment than conventionally grown or raised food. I was also curious about what these other approaches mean—what makes an Eco Apple, for instance—and how they are achieved.

There are a lot of assumptions about what "organic" means, and while it does have a definition, it's hard to understand what the term means in the real world of farming and growing. In the US, it's a certification regulated by the Department of Agriculture, and it specifies a few criteria. For a product to be 100 percent organic, the growers or farmers can't use synthetic pesticides or fertilizers, genetically modified organisms (GMOs), ionizing radiation, or sewage as fertilizer. Meat, dairy, and eggs have to be produced without growth hormones and antibiotics. (Small farms that sell less than $5,000 worth of organic agricultural products don't need to be certified as organic.) They also have to rotate what they grow, plant cover crops (which protect and enrich the soil), and practice conservation tillage (meaning leaving crop residue on the fields before and after growing to reduce erosion). For a product to be labeled simply "organic," it has to use only 95 percent organic substances or ingredients. The remaining 5 percent could be conventional ingredients grown with herbicides or pesticides or other non-organic ingredients.[1]

Here's what it doesn't mean: it doesn't mean that organic growers and farmers can't use any pesticides or fertilizers at all.

They just have to use fertilizers or pesticides that can be found in nature (though some natural chemicals, like arsenic or strychnine, are prohibited). For fertilizer that generally just means manure from livestock; for pesticides it means naturally sourced chemicals.[2] Some scientists say that having food that doesn't have any synthetic pesticides on it is better for you, but studies from the USDA have shown that 99 percent of food produced in the US (from 10,000 samples) had pesticide residues below the limits set by the EPA, and 23 percent had no detectable pesticide residue at all.[3] However, there are pesticides, organic and synthetic, that the USDA does not test for. The EPA, USDA, and many scientists agree that consumers don't need to be too worried about pesticides, though how they affect farmworker health is a different story. It also doesn't convey anything, necessarily, about the nutritional content of the food. Studies have not consistently demonstrated that organic food is any healthier or more nutritious than conventionally grown food.[4] And it doesn't place any restrictions on the size of a farm or the mechanization of it. (For our purposes, it doesn't make a lot of sense to explore this issue, but if you want to know more and haven't read *The Omnivore's Dilemma*, please do.)

There are a few reasons why those criteria can present problems or, at least, can be confusing. We'll start with the pesticides. Pesticides approved for use in organic farming, though not synthetic, are still chemicals. And because they weren't developed as pesticides, they often must be applied in greater amounts than synthetic pesticides in order to be effective. The USDA does not keep records of how much of these organic pesticides are applied, though they do test for residue on food. None of that is to say that all organic farms use toxic pesticides or that organic pesticides are worse than synthetic ones (although at least one study has found that some organic insecticides had a more harmful effect on the species that preyed on the species targeted by the insecticide),[5] just that organic food is not free of pesticides, and the pesticides it does use can have harmful effects as well. Organic farms can also benefit from conventional pesticide use: sometimes organic farms are surrounded or bordered by conventional farms, which use synthetic pesticides, meaning there are fewer pests around to bother the organic food.

In late June (unfortunately for me, in between harvest seasons, so I was not able to get any of the fruit that I need/crave), I went to visit Rogers Orchards in central Connecticut. Over the last two hundred or so years, the family has cobbled together two hundred or so acres of apples, forty acres of peaches, thirty acres of pears, and a handful of acres of apricots, plums, and nectarines, all of which are in a patchwork with homes and other businesses in the area—another reason why it's hard to farm in New England. Across their land, they are trying a variety of techniques to get the best crop they can while using the best science available, and keeping the land as healthy as possible. It seems really difficult.

Peter Rogers, an eighth-generation family farmer who is now one of the ones responsible for the farm, showed me around the orchard in his Jeep, pointing out the new growing techniques he had learned in Holland—dwarf trees grown close together—and varieties of apples the market demands (Honeycrisps, for instance, which we both agreed are overrated for different reasons: for me because they have no taste; for him because they are too finicky to grow). He also pointed out some of the less disruptive, less chemically intense pest management tools: pheromone disruption of certain insects, and white paint on the trunks of young trees to protect them from dogwood bore.

Rogers Orchards uses something called integrated pest management, a philosophy of growing in which pesticides are applied only as a last resort. It encourages growers to employ the most effective option with the least environmental impact and to take action to control pests only when they are poised to pose a problem. It is through the IPM Institute that they received their Eco Apple certification, which requires adherence to these principles and others, and submitting reports of pesticides used, in what amounts, and a justification for each application to Red Tomato, a food hub that manages the certification. It's really expensive to do this: Rogers Orchards has to pay Apple Leaf, an agricultural consulting company, to scout the property to look for certain diseases or pest problems and advise how to proceed: with pesticides, pruning, or a variety of other management practices.[6]

Because they sometimes use pesticides and also can use chemical fertilizers, they can't be certified as organic. But, Michael Biltonen, a scout from Apple Leaf, said it is almost impossible to grow apples organically in New England: there's too much humidity and too much pressure from pests and disease. And climate change is going to make all of it more difficult: weather is less predictable; droughts are more common, as are extreme precipitation events; pests can survive more of the year as the climate warms; new pests from other places will extend their ranges to places they've never been before. Pretending that those challenges can be dealt with without some of the advantages science has provided (fertilizers, pesticides, and other nonchemical tools) is naive, Biltonen said.[7]

In short, to grow the apples that we like to eat and to have them look the way we like, growers have to use some chemical pesticides. Without them, these farms wouldn't survive, especially because growing organically in other parts of the country (Washington State, which produces around 60 percent of the nation's apples and more than 90 percent of its organic apples)[8] is much easier and cheaper. But since the Eco Apple certification doesn't mean as much to consumers as the organic label (if the label is even put on the apples in a store), it's hard to charge the same price for the apples that carry it. Consumers also don't appreciate the difference generally, or how hard it is for some growers in some places to participate in sustainable agriculture, but I think most people in the northeast who buy organic food (who presumably care about sustainable agriculture, or at least I would hope!) would want to support a small, family-owned farm trying to do their best for their family, for the consumer, and for the land, rather than a giant industrialized organic farm on the other side of the country.

All that being said, organically produced food does provide meaningful environmental benefits, and it can be a good shorthand for sustainable (although, it should be said, being certified organic is not the only way to achieve these results, as Rogers Orchards has demonstrated). But the organic movement developed out of a concern for the industrialization of agriculture, which organic proponents thought was destroying soil health, among other natural benefits. As a result, many of the practices associated with the movement and required by organic standards were designed to better (or at least not destroy) the environment.[9] Most of the comparisons and statistics that follow are for organic versus conventional, not organic versus anything and everything, so that would exclude people like the Rogers family. With practices like crop rotation, planting cover crops, and conservation tillage, soil on organically managed land is healthier: it has 7 percent more organic matter, on average, than conventional soil, and is better at holding water.[10] Both of these characteristics make soil less vulnerable to erosion and nutrient leaching when we measure by area (though by product, not so much; I'll get to that). Other practices, like planting barriers of grasses or other species between different crop rows helps increase biodiversity.[11] Organically managed lands are able to support more native grasses and wildflowers, which can help sustain insect, bee, and bird populations. Organic management, on average, leads to a typical increase in organism abundance by 40–50 percent across different populations, and plants and bees benefit the most.[12] However, the biodiversity advantage presented by organic agricultural management is the greatest when the organic land is situated in a landscape otherwise dominated by industrial agriculture and monocultures, because it reduces pest pressure. Call it a study in contrasts. Additionally, water quality—largely because of a lack of synthetic fertilizers—is typically better in organic systems when you look at the area of land. And, because organic farms don't use synthetic fertilizer (which requires a lot of energy to produce), they typically have lower energy inputs than conventional agriculture[13] (though not always), which also means they can be responsible for fewer emissions of carbon dioxide and other greenhouse gases.*

So it sounds like organics are pretty good to the point that you must be thinking that I am making something out of nothing! I am not doing that. There is one major drawback to organic agriculture that affects all of the advantages I just listed: yield. On average, organic agriculture yields around 75–80 percent as much food as conventional agriculture does.[14] Why does that matter? It means that for every acre of organically farmed land, you only get about three-quarters as much food, so to expand organic farming in a way that would still feed the same amount of people, you'd need that much more land. Currently, only around 1 percent of agricultural land around the world is farmed organically,[15] but it is the fastest growing agricultural sector in both North America and Europe; in China, sales of organic produce grew by around 20 percent in 2015, the most recent year data were available.[16]

And as we already know, converting non-cropland to cropland can have important environmental consequences: emissions are

---

* In organic milk, cereal, and pork production, greenhouse gas emissions were higher than conventional; for organic olives, beef, and some crops, they were lower.

released in the process, and that land is no longer as effective a carbon sink as it was. To meet future global demand for food with organics, the world would need to adapt a vegan or vegetarian diet, or else massively expand cropland.[17] And it would be hard to farm according to current organic practices if the whole world were vegan, since organic agriculture depends on livestock for manure for fertilizer (though you probably wouldn't need as much livestock to produce the amount of manure we'd need, especially if the soil were healthier from different growing practices and less fertilizer). If more people were vegan, we would need less land for grazing (since we'd need more lands for growing fruits, vegetables, grains, legumes, etc.), so the intensity of grazing on those lands would be higher, which can have its own negative effects: losing crop cover, poor soil health, increased erosion.

The yield question is also a source of uncertainty—for some crops, the organic yield is pretty close to conventional; for others, there is a significant difference. Some of this has to do with there not being enough nitrogen in these systems to meet the demand of crops, and the cover crops or crop rotation or manure or compost not providing enough nitrogen to satisfy large growing operations. The subject of yield is also one where not using GMOs might be short-sighted: GMO crops, if they are bred to be pest- and disease-resistant, can have better, more consistent yields.[18]

Either way, with substantially lower yields, we can't feed the world with current organic management practices alone. We definitely can't do it if people in other parts of the world start eating meat like Americans do, which they are starting to do. (That's not to say that Americans should keep getting to eat the way we do and no one else should, but that Americans really need to eat less meat.) Plus, as the global population grows, and cities expand to meet this growing population, we will lose more agricultural land to development; climate change will also affect our ability to grow food more generally. To meet future demand with organics, we'd have to clear cut the world's forests and improve technology to increase yield dramatically, or restructure our entire food system.

Some of the environmental impacts are also better or worse depending on how you measure them. If you measure the impacts by area, organics come out on top in terms of nitrogen leaching from soil, water use and quality, greenhouse gas emissions, and some of the other things I just mentioned. When you measure the environmental effects per product (or unit of product, like a pound of potatoes or a bushel of apples), organic products don't always come out on top,[19] or it's just not clear that organic isn't worse than conventional agriculture because of the lower yields and the possibility that more carbon-sequestering forestland would need to be converted to cropland in order to meet future demand—that is, if we all keep wanting to eat organically grown food.

So what should we do instead? Crawl into a dark hole, probably! But also, aspects of conventional agriculture and organic management should be combined, as places like Rogers Orchards have shown. Some of the requirements for organic agriculture might not work in a world with 9.3 billion people to feed. We'll probably still need synthetic fertilizer, although (as we all learned in the chapter about corn monoculture) in much smaller amounts and applied in a smarter way. But farmers would need less of it if they incorporated some of the main principles of the organic movement that we've discussed: crop rotation, cover crops, conservation tillage. These practices help create and maintain soil health and biodiversity; one study found that practicing crop rotation of diverse crops, combined with smaller amounts of synthetic fertilizers and herbicides and a touch of manure, yielded as much or more crop than the normal, nonrotating crops, with standard amounts of fertilizer and herbicide, yields.[20] Plus, the system with less synthetic chemicals resulted in significantly lower water toxicity than the conventional system.

Part of the problem with agriculture is its scale. By making it industrial, we've defied so many of the principles that farmers used over time that made agriculture sustainable before we ever had a need for that word. Industrial agriculture enabled us to feed the world—the percentage of the global population that goes hungry has declined for over a decade (though it rose in 2017 because of conflict- and climate-related stresses)[21]—but one day it won't if we keep going the way we're going. Industrial agriculture isn't sustainable; for a growing global population, which we are, organic agriculture isn't either. Barring some scientific discovery that I can't imagine, finding something in between is most likely our best bet.

But does it matter how our food is grown if we're constantly shipping it all over the world?

How Far Our Food Goes

Walking into a grocery store in 2019 and picking up a fruit or vegetable at random is not totally unlike spinning a globe to see where your finger lands. It could be home, it could be somewhere on the other side of the world, it could be your neighbor to the north or south. Food comes to our grocery stores from everywhere, all the time.

It's not surprising that food moves around—exchanging spices and foods was the lifeblood of empires, from ancient Rome to the Columbian exchange. So maybe, given how much easier it is to move things now than it was in either of those times, it's also not surprising that different foods grown in different parts of the world could end up in a random grocery store almost anywhere in the country.

But the ease of movement of all of this food and the quantity of it—pyramids of lemons, towers of squash—make it all seem normal that this should happen, that it must make sense somehow, or otherwise it wouldn't happen. It's a sign of progress, perhaps, that we can eat a turnip in the summer and a strawberry in the winter and that that is life in 2019.

I started thinking about the way food moves around because I wanted to include something in this book about food miles, or the carbon footprint embedded in a certain food based on how far it travels from farm or factory to fork. Part of that comes from my own sense of how things have changed even during my short, precocious, wildly achieving lifetime: you used to have to wait until June to get fresh berries, we used to eat asparagus only in spring and apples only in the fall, and you couldn't find a cherry before summer or after it had cooled into autumn. That's not really true anymore: an apple or a raspberry is undoubtedly better in its season, but I can get it whenever I want, and I can easily get things that were exotic not that long ago: avocadoes, mangoes, kiwis, etc. I imagined that there must be an environmental impact associated with that change, and also I thought that the whole thing seemed crazy even though I like it. Both things are true, but not necessarily in the way that I thought.

First, I want to set the stage—a spotlight, me in a black turtleneck, slowly applying ChapStick while sitting on a stool—by explaining the current scope of the always moving feasts we eat. It all gives a whole new meaning to Go-Gurt. That will make it easier to understand the environmental impact of all that moving around and whether shipping delicate and perishable produce, meat, dairy, and eggs around the world is as wasteful as it sounds.

Around 15 percent of all the food in the US is imported from another country (we are still a net exporter of food, though most of that is made up of crops like corn, soybeans, and wheat).[1] But what is imported is the most interesting part: more than half of the fruit we eat (compared to less than a quarter in 1975) and 20 percent of the vegetables come from abroad. And it's not just more of the same stuff we always ate: the market has completely changed, too—over that same period (1975–2017), per capita mango consumption, just as an example, increased by 1,850 percent.[2]

First of all, what?! Second of all, why? There are a few reasons. First, agricultural scientists have developed new varieties of crops that can succeed in non-native climates, and they and others (farmers, growers) have figured out how to make crops do better in those places and elsewhere. Second, as the American population (though this is not unique to America) got wealthier, people wanted different things, namely more fresh produce and less of the processed, packaged food that had once seemed totally futuristic and now seems like it might kill us: Twinkies, Spam, Wonder Bread. Third, immigrant groups from all over the world brought new foods or ways to use them to their new homes that have become totally mainstream (put an avocado on it!). Plus, globalization. Those things all add up, but the rapid, skyscraper growth of food imports is largely due to a massive regulatory overhaul in the US that allowed in fruits and vegetables from abroad that previously weren't allowed because of fears (real ones) about diseases and pests. But the ways of containing those things (and the fears) changed, and so did the approach to importing food. Now we have lemons and cherimoya (which Mark Twain apparently called "deliciousness itself")[3] from Chile; peppers from Peru; persimmons from Japan; raspberries, blackberries, and pitaya (aka dragon fruit) from Ecuador; avocadoes from Colombia; and thousands of other fruits from other places. And when it comes to some more common tropical fruits, like bananas, limes, and pineapples, 99.9 percent come from somewhere else. And if you're wondering whence they hail, the answer is probably Mexico: 46.2 percent of our fruit comes from there. (Chile sends about 15 percent, Guatemala 9 percent, and Costa Rica about 8 percent.)[4]

And there's not necessarily anything wrong with that. In fact, there are a lot of advantages to importing food: jobs in developing countries, some environmental benefits, and more interesting, more nutritious food for us. But it also means that some American crops (and therefore American people who grow, tend to, and harvest them) suffer: we don't eat as many peaches, cabbages, celery, or oranges as we used to (because we like the new stuff better), and most of those sold in the US are grown here, too.[5]

Food is moving around in quantities and at speeds that would have astounded the ancient Romans and maybe even your grand-parents. Now let's put that in its environmental context. (If you're wondering why there isn't anything here about the environmental impact of shipping, the primary way food moves around, we'll get to it in another chapter.)

About ten years ago, we all heard a lot about local food and locavores and how much better they were than the rest of us, and that we should be shopping at farmers markets and not waste fuel buying food that was trucked from far away when we could have gotten it from a nearby farm. It was hard not to feel like we should all cut those people out of our lives. Now, it seems like people talk about that less, or they talk to me about that less, or "people" talk to "me" less in general. A term that was used in those con-versations was *food miles*, which originated in a 1994 report from a British advocacy group on the energy and resource efficiency of moving food around.[6] As a result, a lot of the research about the issue comes from Great Britain, which is slightly different from most other places because it's an island, but the same principles apply to the US.

That report and other scientific studies have tried to answer this question: is it more efficient to grow food in one country and sell it in another than to grow and sell food in the same place? The assumption in asking that question is that adding extra legs onto a food's journey would unnecessarily add greenhouse gas emissions from transportation to the atmosphere.

As it turns out, the emissions from transportation relative to the overall emissions associated with food production are very small. All transportation (including how farm equipment or other supplies got to the farm, for instance) is about 11 percent of the total, and final delivery (from producer to retail location) is about 4 percent. Generally speaking, the production of food accounts for much more of the emissions as well as the overall environmental impact. For beef, which, as we know, requires tremendous amounts of energy and resources to produce, only 6 percent of greenhouse gas emissions are from transportation (1 percent is food miles). For fruits and vegetables, transportation's share is higher—about 18 percent—but still less than production. What does that tell us? It means that buying local doesn't make that much of a difference to food's overall impact. One study found that you could reduce your emissions by a maximum of 5 percent if you bought local. If, instead, you shifted one day's consumption of red meat and dairy per week to another protein source or fruits and vegetables, you could achieve the same level of greenhouse gas reduction.[7]

But there are other reasons why food miles are not the all-important metric that people thought. Some of that is because of things like climates and growing seasons around the world, but it can be harder to grow food on overextended soil, like in the US and Europe. Here are some interesting examples that help illustrate that phenomenon through contrast. A study from New Zealand showed that it was more energy-intensive to grow apples, raise lambs, and produce dairy in the UK for year-round British consumption than to produce those same things in New Zealand and ship them to the UK periodically, a journey of more than 11,000 miles. How can that be? In the UK, apples grow only at a certain time of year and would need to be stored (and refrigerated) for the rest of the year—about six months of refrigeration—which uses a lot of energy.[8] As for raising the animals and producing the meat and dairy, New Zealand's ample space, few people, sunshine, and climate means that animals can eat grass mostly (instead of feed) and live outside, requiring less energy than keeping them in a heated, lit barn, like they probably would be in the UK.[9] However, if British apples were eaten only in season and not refrigerated, that would be the best option, but our expectations have changed, and now we need a constant stream of apples rolling into our open mouths all day long. The same was true for Spanish tomatoes versus British greenhouse-grown tomatoes, since the greenhouses were heated by fossil fuels (in this study), despite the lower yield of the Spanish tomatoes (they were probably taking a siesta and eating dinner at 10:00 p.m.).[10]

To generalize, growing food in places like New Zealand or in some parts of the developing world means lower inputs of energy: less synthetic fertilizer, less mechanized farm equipment, less refrigeration (though cargo ships are temperature-controlled).[11] By contrast, year-round domestic production in much of the US and Europe would require greenhouses, likely powered by fossil fuels, since not every country has a California or a Florida or a Texas where food can be grown all year, though even those places have growing seasons. But we may not always be able to count on those places, especially as climate change makes extreme weather events more likely and more intense. In 2017, the US experienced sixteen different major environmental disasters or extreme weather events whose costs each exceeded $1 billion, which in total have been estimated to cost more than $5 billion to agriculture alone.[12] Plus, food grown locally or domestically still has to travel, likely by truck, at some point, and most people are driving their cars to the grocery store. Studies have also shown that growing in at least some parts of the Southern Hemisphere is more efficient in terms of water use and prevents nutrient pollution (and therefore eutrophication), since less fertilizer is needed because the soil is less depleted and can provide nutrients naturally to the plants.[13] Not to mention that the conversation about food miles mostly preoccupies itself with produce and where specific ingredients came from without thinking about the larger supply chain. We don't really ask how every ingredient in a flavored yogurt got to the factory where it all came together.[14]

Even organic meat might eat organically grown feed from somewhere else, or the manure to grow that feed came from far away. For conventional produce and meat, the chain could be even longer. So, it's hard to account for the supply chain, and it's very easy to starve to death while trying to determine the exact provenance of every single item that was involved in producing what you eat before you eat it.

If you aren't eating everything in season and come from the US, the UK, or somewhere similar, it is not necessarily more energy-intensive to eat food from abroad, and in some cases, it might be more efficient. But as we know, emissions from energy use are not the whole story, and if you are buying from nearby small farms and eating food when it's in season, your overall impact on the environment is probably less damaging—but you have to be really good at food shopping to make that matter. In the end, though, buying local or not isn't going to save the planet, and within the current food system, it could, perversely, grow your carbon footprint. However, that conclusion—if I were to leave it at that, which I absolutely cannot do—does two things.

First, it accepts the current energy-intensive, industrialized agricultural system, and it advocates outsourcing our emissions to other countries. Since we're not counting those emissions on our own balance sheet, we are exporting our emissions by growing food in Mexico, which is a quirk of accounting rather than actually meaningfully reducing our own emissions. At scale and over time (as the population grows and as more people can afford and want fresh food), this practice could do to other countries what large-scale agriculture has done to us: deplete soil nutrients, drain and pollute aquifers, blah blah blah, you get it.

Second, it obscures the dizzying amount of food and everything else that we ship around the world because we can, and because it's become cheaper to grow food in one place and ship it to another. At the beginning of this chapter, I explained how much food is imported and how that number has grown like crazy. In 1999, we imported 48 million tons of vegetables; in 2017, we received 111.6 million tons of asparagus, tomatoes, and more, burning dirty fossil fuels, mostly on massive cargo ships, polluting the water, releasing greenhouse gases, and adding to ocean acidification, and doing the same with trucks. (And that doesn't even capture the vast amount of food that goes from California to just about everywhere else in the country by heavy-duty freight trucks.)

So food miles might not be a problem in and of themselves, but they are a symptom of larger ones in the food and transportation systems, in which the consequences are externalized, so we act as though they don't exist. No one is paying for the costs right now, but we will eventually be getting the bill and it will cost a lot—and more than money.

There are few areas where our ignorance about where things come from and where they go may matter more, in the end, than in the oceans.

A Sea of Troubles

Of all the things I've written about so far, writing about fish has been among the most challenging. Is it because I think fish are kind of gross? Probably. Is it because I don't really like to eat fish, and I don't think that most people do, either, but they pretend because they want to seem better than me? Also probably yes that is true. All because they "want to eat healthy"? Sure, fine. But without making this about me (why stop now?), it's also because writing about fish is possibly even more complicated than some aspects of the food system I've discussed up until now—it involves the wild in a way the other areas don't, and it also involves industry, mainly in the form of aquaculture (seafood farming). Commercial fishing and aquaculture are both deeply connected to aspects of the food system we have been exploring in some ways, and entirely separate from them in others. Fishing and aquaculture involve international agreements, regional regulations, natural fluctuations, the food chain, climate change, and pollution; they also involve corn (duh), antibiotics, pesticides, global trade, the destruction of native ecosystems, and loss of biodiversity. The two ways of getting fish are interconnected, even as one presents itself as a sort of substitute for the other.

Since the other issues I've written about so far in this section go together—they're all about growing food or what happens to it after it's grown—why did I bother writing about fish, which is pretty different? Well, it's important to me personally, and I think important to discussions about the future of human and environmental health. As I said, I don't like/am afraid of fish because they are slippery and therefore not to be trusted, but I am fully in awe of the oceans. Without them, the planet would already be on fire; they are mysterious and unknowable; it feels good to swim.

It's because of the ocean that I wanted to write about the environment in the first place. When I was studying history as a graduate student, I read *The Mortal Sea: Fishing the Atlantic in the Age of Sail*, by W. Jeffrey Bolster. Reading about centuries of exploitation of the ocean combined with general human ignorance made me realize that while climate change may be an especially intense and new problem, we have been dealing with versions of it for as long as there have been people, and that we take life and the planet for granted, but it is always changing and responding, often to the things we do, even if we are unaware of it. That book is dense and full of statistical information about historical fisheries, which might not be for everyone, but it is also vivid and urgent. I wanted to keep telling that story, somehow.

I could write a whole book about fish and the oceans, and there are plenty of books on those subjects. Here, I hope I can provide a brief but smart and helpful framework to understand what is going on in the ocean. There are lots of ways I could do that, but the two most interesting ways to me are to talk about how climate change is affecting global fisheries and small finfish, like anchovies. (It will make sense why.)

Around the world, different countries are handling the challenge of how to fish the oceans sustainably in different ways. It's a big challenge: seafood (including fish, crustaceans, and mollusks) accounted for almost one-fifth of the animal protein the global population ate in 2015; more than 60 million people support themselves by working in either fishing or aquaculture.[1] There are international waters where there are virtually no rules about what can be fished and how much, and even if there are regulations, hundreds of miles from any coast, lots of fishermen from many countries just don't pay attention. A 2009 study estimated that one out of every five fish purchased by an American was caught illegally,[2] and we import about 90 percent of the fish we eat.[3] Some environmental advocates argue that more than one in five are caught illegally, since other countries don't abide by our rules about how to fish to avoid catching or harming marine mammal and sea turtle populations.[4] Historically, we also haven't paid attention to how ecosystems and populations have changed over time or how they interact with each other, or we've just ignored the changes because they got in the way of what we liked to eat or how much money we could make. An example: Americans fished the Atlantic salmon population with such aggression that, in 1948, the commercial fisheries closed down, and both commercial and recreational fishing of wild run Atlantic salmon remain closed today, by order of the federal government. There used to be salmon in every river north of the Hudson. Now, they're in only about eight rivers in Maine, and they number in the few thousands.[5]

We didn't learn the lessons from that until after many more species were on the brink of disappearing. In the late 1990s, things were bleak: a record ninety-eight fish species were considered overfished in the US, and several New England fisheries had collapsed, including cod, flounder, and haddock.[6] The federal government stepped in and mandated that rebuilding plans based on science be implemented for every species that was considered overfished. In US waters today, only thirty species are considered overfished, and forty-four species have been rebuilt.[7]

Globally, things look different. Marine fisheries peaked in the 1990s and haven't recovered; management is not great around the world. We are fishing 55 percent of the world's oceans,[8] about one-third of global species are overfished or fished unsustainably, and nearly 60 percent of stocks are considered fully fished.[9]

But management is just one problem facing fisheries. The other one, which we talk about even less than overfishing, is how greenhouse gas emissions are changing the oceans. We take the oceans for granted. (Since most of the earth is ocean, our planet was, perhaps, given the wrong name.) The ocean is alternately a resource frontier, a garbage dump, and a global transportation platform, but also a source of life, artistic inspiration, and renewal. But it may be that we are the most ungrateful for the oceans when it comes to climate change. There are a few areas where this is made manifest—coral reefs, New England cod—but, largely, we are unaware of the scope of the problem. Since the 1950s, the oceans have absorbed more than 90 percent of the excess heat created by greenhouse gas emissions; if they hadn't, the earth could have warmed by an average of 36°C[10], instead of 1°C or so.

As a result of absorbing all that heat, the sea surface temperature is warming, which is strengthening hurricanes, melting the polar ice caps faster, and accelerating sea level rise. Temperature shifts also have dramatic consequences for world fisheries. Already, scientists, fishermen, and others are noticing that fish populations are migrating toward the poles or farther offshore, looking for cooler waters.[11] In these warmer temperatures, fish may not be able to reproduce or grow at normal rates.[12] And those complications are expected to increase: a study that mapped projected migrations in North America for 686 fish species under various emissions scenarios anticipates that most species will move outside of their historic habitats, confounding international and national management strategies and making sustainable fishing even more complicated and difficult than it already is.[13] Warming also causes more frequent bleaching of most coral reefs around the world, and they have little chance for recovery.[14] Coral reefs are really important. Even though they cover less than 1 percent of the ocean floor, they support about 25 percent of all fish species.[15]

A really good example of how climate change has already affected fish and might continue to in the future is the New England cod. For a little bit of historical background (because who can resist stories of cod from days gone by), Captain John Smith wrote during a visit to New England's waters, "He is a very bad fisher, [who] cannot kill in one day with his hooke and line, one, two, three hundred Cods..."[16] Three hundred! In a day! While that amount of fish is a personal hellscape, it is staggering to imagine now: on the Georges Bank, which stretches from Newfoundland to Massachusetts, once one of the world's most productive fishing grounds, generally and for cod specifically, 9,600 square kilometers have been closed to cod fishing since 1994, when it was discovered that the cod stock had declined by 40 percent over the previous four years.[17] And still, cod aren't recovering: the Gulf of Maine, separated from the rest of the Atlantic by the Georges Bank, warmed by 0.03°C every year from 1982 to 2004, three times the global mean rate. Beginning in 2004, the warming rate accelerated sevenfold, peaking in 2012. The Gulf of Maine has warmed faster than any other part of the global ocean except the Arctic; without climate change, the likelihood of that warming taking place is 0.03 percent.[18] A result of increasing greenhouse gas emissions, this warming is likely worsened by shifts in ocean currents as well. Commercial cod fishing in the Gulf of Maine is allowed, but despite strict quotas determined by the best science, the population isn't recovering.[19] Curious as to why, scientists discovered that warming ocean temperatures, especially in summer, were affecting cod's ability to reproduce and causing a decline in the number of fish reaching maturity.[20]

Acidification is another problem: all the carbon dioxide absorbed by the oceans has changed their chemistry. Absorbing that much carbon dioxide turns the oceans more acidic, which makes it harder for mollusks, crustaceans, and corals to grow their shells and skeletons, because the chemical composition of the water can dissolve them.[21] In more acidic conditions, harmful algal species, which can sicken fish, marine mammals, and people, may proliferate even more than they already do, most commonly a result of nutrient pollution.[22]

Deoxygenation is yet another problem. Rising temperatures and the changing chemical composition of the ocean (the acidification from carbon dioxide as well as nutrient pollution from fertilizers, etc.) has caused the oxygen levels in many areas of the ocean to decline. Sites of minimum levels of oxygen in the open ocean have grown by several million square kilometers,[23] and coastal areas have low enough concentrations of oxygen to prevent marine species from reproducing, surviving, or migrating as they used to.[24]

Besides reducing our emissions, scientists say we need to better measure all of these effects so we can get a better sense of what has already happened in the oceans and how they might change. They also recommend that 30 percent of the oceans be protected as marine reserves to protect dwindling fish stocks and allow them to recover. Instead, the Trump administration wants to keep the marine protected areas we have but open them up to fishing, which seems like the opposite of protecting them.[25] Currently, only about 3 percent of the ocean is protected.[26]

There is another story I want to tell, and this one is about aquaculture. In 2019, the global population is expected to eat more farmed than wild fish for the first time.[27] Clearly, we need aquaculture: seafood, including farmed seafood, account for about a fifth of the global population's intake of protein. As the global population grows we'll need more food, including more fish. There are lots of problems with aquaculture: antibiotics, contamination, and pesticides are some of the ones you've likely heard of, because they can affect human health. Aquaculture also has a huge environmental impact: farming catfish has as big of an environmental impact as industrial beef production, according to a recent study that compared different types of animal protein production. Overall, livestock production used less energy than some forms of aquaculture, specifically catfish, shrimp, and tilapia. The way to have the smallest environmental impact (even smaller than a plant-based diet) would be to eat farmed mollusks and small wild fish.[28]

But I want to talk about how aquaculture relates to overfishing. Some farmed animals, particularly farmed bivalves (clams, mussels, oysters), can live on plants (phytoplankton), and so can tilapia (though they can also be fed on corn, which complicates the whole thing). Salmon, one of the most popular fish in the US, cannot. Once salmon reach maturity, it's thought that they need to eat other fish. Usually, it comes to them in the form of meal and oil derived from anchovies or similar species. These species are crucial to the ocean's food web, but they also represent the world's largest fishery: these small finfish are 37 percent of all fish caught, up from 8 percent 50 years ago.[29]

People can eat these fish, but we mostly don't. Instead, they are ground up and processed and made into meal and oil to be fed to salmon or other land-based livestock, or made into supplements that we take or add to things like milk and eggs. (Many scientific studies show that there are no conclusive health benefits to taking fish oil, which is rich in omega-3s, through supplements.)[30] The growth of aquaculture and the supplement industry is having a huge effect on these types of fish: in 2010, menhaden (a small, silvery fish native to the northwestern Atlantic) had been reduced to 80 percent of its maximum potential, with 80 percent used to make meal and oil.[31] If these trends in aquaculture and agriculture continue, these wild populations will likely be overfished by 2050 or even sooner, even if they are harvested sustainably in the mean-time.[32] Between 60 and 70 percent of the 15 million tons of wild fish caught are used to make meal and oil.[33]

The world's largest producer of farmed fish is China, which is responsible for much of the fishing of these types of small fish, especially off the coast of West Africa.[34] In several West African countries, Chinese companies, which often buy licenses to fish in these countries' waters, have forced local fishermen out of the game, since they are able to scoop up in a day what these fishermen would haul in a week; they have built dozens of fishmeal processing plants up and down the West African coast, and the product is brought back to China to feed the booming aquaculture industry.[35] A lot of the farmed fish stays in China, which consumes about one-third of all fish worldwide.[36] But China also exports to the US, as do several other Asian countries. Ninety percent of the fish consumed in the US is imported; meanwhile, we export about one-third of all the fish we catch.[37]

What is that about? Mostly, Americans don't like fish that tastes like fish, so we seem to prefer farmed, tasteless fish like tilapia, farmed salmon, catfish, and shrimp to the fishy fish that are relatively abundant in some US waters, particularly Alaska.[38] But more than that, it tells us that this is a problem everywhere. Sure, the US is managing its fisheries better than the average, but China and lots of other countries aren't. If we eat farmed fish from those places or fish caught by boats from those countries, we are participating in destructive practices around the world. Just like the rest of the food world, it's all part of the same system. There are ways to act more responsibly. One way is to look for sustainability certifications from third parties. The Monterey Bay Aquarium and the Marine Conservation Society both have smartphone applications that can help you figure out if the fish you're about to buy is problematic, though they don't include all species.

There are more problems to come. So far, we have managed to avoid in the sea what we have done on land, namely, cause hundreds of species to go extinct. In the oceans, as far as we know, only fifteen species have gone extinct in the past 514 years, and none in the last five decades (that we know of), compared with five hundred on land.[39] Much of those terrestrial extinctions came after the Industrial Revolution had taken hold, which is important to keep in mind. According to Douglas McCauley, coolest marine biologist, winner of Vice's "Human of the Year"[40] designation, poster of his Fitbit results and his genotype on his academic website,[41] who doesn't eat fish he doesn't catch, told me that we are on the brink of an industrial revolution in the ocean.[42] As we race to deplete finite resources on land—precious metals, rare earths, fossil fuels—some companies are planning to establish those industries in the ocean. There is already some ocean mining and ocean drilling for gas and oil, which has gone super well so far: just look at great success stories like the Deepwater Horizon spill in the Gulf of Mexico, as an example. (In case it's not clear, I am being very, very sarcastic.) On the massive industrial scale that could take hold, we don't know what the effects could be. Sediment plumes from industrial activity underwater could have serious negative consequences for marine populations; we already know that the noise from underwater drilling and seismic testing has harmed marine mammals.[43]

"When you actually look at the glass half empty, the vulnerability for these heavily impacted species, well, we could easily engineer a mass extinction in the ocean," McCauley said. "We could easily lose our capacity to rebuild. We have an immense amount of opportunity, but we're poised to lose it. This is kind of the make-it-or-break-it generation."[44]

Despite the grim future McCauley is predicting, he is more hopeful than me and what those quotes suggest. He hopes that

Industrial Revolution: The Sequel will go better than the one before when it comes to the environment. "We have the chance to start the Industrial Revolution all over again with a heightened awareness about how to manage the environment," he said. "We could make it go right."[45] Fingers crossed.

If there's no food, there's no us. Few things to me are as short-sighted as our disregard for the environment in order to feed ourselves, as if we won't continue to need clean water and healthy soil and pollinating insects if we want to keep eating. The relentless plunder of the oceans isn't actually all that different from the way we treat the other natural resources that feed us—as if they are limitless and eternal and unaffected by how poorly we treat them. They're not.

Fashion

The fashion industry does not reveal itself easily to me, and that's not because I have bad taste, although you can never be sure. Few companies, both outside the fashion world and particularly inside of it, seem particularly eager to share the environmental effects of what they do because usually they are bad. But fashion seems to have gotten away with a lot—serious and often damaging environmental effects, and a complete lack of transparency. They provide almost no information about what they do and what they're trying to do to be better, if anything.

Really, there is not a lot of data, which has made writing about the science of fashion and its accompanying effects particularly frustrating.

I have some theories about why this is true. Fashion has traditionally and stereotypically been seen as something that women are interested in, and many things that women are interested in are assumed to be frivolous—certainly not worthy of serious scientific inquiry, for the most part. Consider this my attempt to topple the patriarchy.

Meanwhile, fashion is a $2–3 trillion industry that literally touches almost everyone on earth. And our vanity, plus the ubiquity, necessity, and lack of transparency of fashion seem to make us more complacent about its production, especially as clothing gets cheaper. These things combine to make us more willing to accept things we know can't be great for the planet, if we take a bit longer to think about it.

The problems with that current approach are many, and really serious. Linda Greer, a former senior scientist at the Natural Resources Defense Council, helped me figure out how to think about the fashion industry; she's one of the few people who does, in a systematic, scientific, and global way. One of the most interesting things she said was that "the manufacturing of our clothing is so fundamentally harmful and irresponsible" and yet clothing companies expect the consumer to understand and demand different and better practices if they'd like to see changes. "The pharmaceutical industry doesn't say, 'We'll make safe drugs if our customers demand it,' nor do food companies," she said. "There is a sense on the part of most of these companies that the consumer needs to demand what they want, and then we can shop our way out of this problem." However, they don't provide the information that a consumer would need to be able to demand something different—something responsible—and even experts like Linda often can't figure out how to make the right choices: "I know way, way more than most people would want to learn," she said, "and it still doesn't help me in my own ability to reduce my footprint." She also told me that Stella McCartney, a fashion brand and designer committed to sustainability, has at times tried to figure out how to source a certain product in a sustainable way, but the employees discover they can't find any information and end up commissioning reports and studies themselves (see: viscose rayon), which isn't really how this should work. Producers and designers and retailers and customers should all be communicating about the materials and their impact.[1]

It's unrealistic to expect consumers to demand responsible production because we, the consumers, don't really know how things are made. A lot of the time, we have no curiosity or we are afraid to find out since we have a sneaking suspicion it's not great. True, most clothing companies have some kind of sustainability page on their website. Unfortunately, these pages usually contain little or no specific information about how the company's products are actually made. Sometimes, they sort of abandon responsibility by saying that they try to ensure "transparency" in their supply chain, but give no actual evidence of doing that or an explanation of what it might look like. According to Greer (and confirmed by me), these sustainability pages are almost always just PR pages, and the communications departments at these companies—even the ones that actively advertise their sustainability and don't just put up a site because they have to—are almost uniformly unresponsive or point you back to the website. And then they say that consumers don't demand sustainability or responsibility, but how would they know?

We, the consumers, could do more, to be sure. Hopefully, reading this book will give you a better sense of the kinds of effects the fashion industry has on the environment. You'll notice that those I've written about here are pretty specific, focusing on a particular aspect of pollution, rather than the overall effects.

That's partly because of the lack of data that I mentioned before. Even Greer said she sometimes can't find information. But if you do look for information about the environmental impact of the fashion industry, you'll probably come across two major claims: the fashion industry is the second-most-polluting industry in the world,[2] and it's responsible for 10 percent of greenhouse gas emissions.[3] The problem with the first claim is that it's not true, or at least there is no source to prove that it is true. Apparently, there was a study from the Danish Fashion Institute that originally offered this claim, but they have since walked it back.[4] The problem with the second claim is that there are lots and lots of different estimates about the precise greenhouse gas impact of the fashion industry. This may be because it's an industry that intersects with so many others—agriculture, fossil fuels, transportation, chemicals—so it's difficult to attribute effects to fashion specifically. Additionally, much of the production happens in places with lax climate accounting and environmental standards, which makes its emissions even more difficult to account for.

But I don't mean to suggest that because those two things aren't true or are imprecisely offered, we don't need to worry about fashion. The fashion industry is huge, and it's everywhere, and it interacts with everything: animal agriculture, plant agriculture, petrochemicals, mining, construction, shipping, and manufacturing. It's huge! It contains multitudes!

The trap with writing about the overall fashion industry is partly that it's challenging to write about it comprehensively for the reasons I've explained above, and partly because it's so big that trying to do some big-picture overview would be unsatisfying. So I decided to write about a few materials and explore specific ways that they create pollution or have a damaging effect on the environment.

So when I wanted to write about cotton, I decided to write about denim, because it's a particularly American thing and a particularly water-intensive thing to produce, and it illustrates, in a vivid way, the general problems of water consumption in agriculture and how wasteful we are with our water supply in an industrial context.

I wanted to write about athleisure because I think it's one of the weirdest developments in my lifetime. Don't get me wrong— it's comfortable, and it wicks my sweat away. (But always remember, the sweat has to go somewhere.) But mostly, when I'm talking about athleisure, I'm talking about synthetic fibers. Because we're going to talk about fossil fuels in a more precise way later on, I wanted to talk about a particularly insidious way that synthetic fibers contribute to pollution: microfiber plastics. They're everywhere, and they're there forever, and there's, so far, nothing to be done.

I'm also fascinated by fast fashion and the companies that have transformed the industry overall by producing cheap things as quickly as they can, and doing so with almost no creativity or integrity by ripping off high(er) fashion designers. There's something appealing about being able to buy really cheap, fashionable clothing—duh—but it has given us a false sense of inexpensiveness. It's not only that the clothes are cheap; it's that no one is paying for the long-term costs of the waste we create just from buying as much as we can afford (or think we can afford).

I wasn't going to write about viscose rayon, but Linda Greer told me I should, and it turned out she was right. It's marketed as a natural fiber, but the chemical usage to produce it is so intense that it shouldn't be considered natural at all. And where does it come from? Primarily from chopping down forests, that's where. To me, that seems like a bad way to produce a natural fiber. In the next section, we'll get to the more specific impacts of deforestation when we look at biomass, but there are some specific geographical problems from rayon production, and I just need you to know them.

Another thing fast fashion has helped create besides waste is the phenomenon of cheap cashmere. I wondered how we got to a point where one of the ultimate luxury items became very accessible. It's not because we figured out how to make it in a lab or breed goats to have four times their natural amount of hair. It's because we are creating a demand for soft, cheap sweaters that is stretching the high mountain prairies of China and Mongolia to their limits, shooting ourselves in the foot and the goats in the hooves to get it.

It's not as if these are the five things—denim, athleisure, fast fashion, viscose rayon, and cheap cashmere—that make up the fashion industry and there's nothing else. There are accessories—shoes, bags, jewelry. There are animals that are treated cruelly so that we can wear their skins and furs as coats or hats or shoes; raising cows and other livestock for leather or wool also adds significantly to greenhouse gas emissions, as we know from our section on food. There are millions and millions of people who work in terrible conditions for very little money and are exposed, daily, to truly harmful chemicals. And there is the beauty industry, which is also out of control and underregulated and subject to the oiliest of snake-oil marketing techniques. Lots of cosmetic products use petroleum-based chemicals, and in addition to being made of fossil fuels, those chemicals can emit as much air pollution (in the form of volatile organic compounds, which are smog-forming) as cars, according to a 2018 study.[5] Cosmetics also use a lot of water (often, it's the first ingredient); the waste chemicals used in our shampoos and soaps (which we also need water to wash off) can be toxic, and those toxic chemicals, washed down the drain, may get into our water systems. And some beauty companies still test their products on animals. In the last few years, it seems like there has been a lot of focus on skincare and self-care, and I'm curious as to why we want to live forever with perfect skin at the precise moment that the world is ending...?

These five things give a sense of the wide range of impacts associated with the fashion industry. It's an industry that relies on limited physical and natural resources, and we treat it like it has no limits at all. We create waste based on the creative impulses of a magazine editor or a fashion designer; we throw away things that are perfectly good and usable and took energy and resources to make because someone tells us it's not cool anymore. More than any other area, it's subject to the fickle winds of human desire and aesthetics. Even if we think it has nothing to do with us—we're just wearing our sneakers and jeans and T-shirts, year in and year out—it does.

As Yves Saint Laurent or maybe Coco Chanel once said, "Fashion fades, but style is eternal." As I say, "Also, plastic. Plastic is forever." (You will soon find out why that's relevant. Cliffhanger!)

Thirsty for Denim

The ubiquity of denim is one reason to write about it: it's familiar, everyone knows what it is, everyone can see themselves in its story. The other reason, which is not unrelated, is that denim is a peculiarly American thing. Many of us wear jeans every day; we associate them with leisure and comfort and casualness, even as they have been transformed into luxury items from expensive fashion houses, selling for hundreds of dollars if not more. There's that part of it, which is American, even though the particular weaving of denim is European, possibly English or French. But in the American context, denim grew up with America. It was there, and even critical, during expansion across the continent, linked with a singularly extractive economy and built on the backs of enslaved people and the genocide of Indigenous peoples. That may sound hyperbolic for a simple pair of pants, but it's true—or I'm going to argue that it is.

There is some evidence that working people—sailors, barbers, seamstresses—had worn denim in Europe before it ever became associated with laborers in the United States, but I'm not going to give you a full history of denim even though I could. Believe me, I could.* Instead, I'm going to start our story with Jacob Davis and Levi Strauss, whom you may have heard of. Jacob Davis was a

---

* Another option would be to read *Making Jeans Green: Linking Sustainability, Business and Fashion*, by Paulina Szmydke-Cacciapalle.

tailor in California, who in 1870 received a request from the wife of a woodcutter, whom she described as "bloated with dropsy," to fashion him a pair of durable pants he could wear while working. (Like most straight married men, he needed his wife to do everything for him. Burn!) Davis got some fabric from Levi Strauss, a Jewish immigrant and dry goods wholesaler in California, who had gone west in 1853 to make his fortune outfitting all those also seeking their fortunes in the California gold rush of the late 1840 and 1850s. Soon, Davis's design was all the rage, and in order to fend off imitators, he decided to apply for a patent. But he couldn't afford the whole cost, so he enlisted Strauss's help for a 50-50 split of the profits. Davis sent two samples—the original white cotton duck he had made for the woodcutter, and a blue denim fabric from a textile mill in New Hampshire, which Strauss already stocked in his shop, so blah blah blah we're all wearing blue jeans 150 years later. The blue jeans were quickly adopted by gold miners in California, who spent a lot of time kneeling in streams, panning for gold, and liked the pants' durability. It's at least in part because of the gold rush that California was settled by such a wide variety of American and non-American migrants (many Mexicans were already living there). But it's also the reason that many native communities were destroyed: "All the mining strikes in the West were an unmitigated disaster for Indians," and not just in California.[1]

I'm not putting all of that—genocide, the beginnings of an extractive economy—on the backs of Jacob Davis and Levi Strauss, or on the humble blue jean. But I do want to complicate the narrative of the blue jean as an uncomplicated American symbol. We also associate jeans with cowboys settling the West and taming the frontier (largely a result of marketing by Levi's and others as well as Western movies and novels), but a large part of that "taming" also included massacres of Native Americans, and I think it's important to fully confront our national myths, even when it doesn't seem to bear directly on the modern-day environmental and climate impact of denim production, because it does. We live in a society that gained much of its wealth from the extraction and exploitation of natural resources, and the denim industry (and the fashion industry more broadly) is no exception to that. In that way, denim is essentially American: it's built on an industry that was built on environmental exploitation, too. Jeans have, of course, taken on a life since then—a symbol of the working man (*blue collar*, from a denim shirt) and of the counterculture in the 1960s and '70s—but I think it's important to think about how that all started, because it also spread to the rest of the world. And it has spread: the denim industry produces over 3 billion denim items every year, and we spend $93 billion buying them.

But now for the jeans themselves. Talking about jeans is, principally, a way to talk about cotton, which is also a product whose ubiquity and importance is inseparable from the labor of enslaved persons, mainly in the American South, and another way to connect the American economy and its accompanying environmental degradation to racial injustice. A lot of what follows applies to most things made out of cotton, not just jeans. It makes sense that cotton and anything made from it (like denim) would affect the environment: it's made of something that grows from the ground, and as we've seen from learning about food, that necessarily affects the natural world. As we also learned from the food industry, we don't always grow, produce, and transport things in the most efficient or environmentally friendly way.

Cotton is, as the elders have taught us, the fabric of our lives. It's grown in more than sixty-five countries around the world, makes up about one-third of all the fibers used in textiles, takes up about 3 percent of global agricultural land,[2] and has a big carbon footprint: producing the world's cotton supply for use in textiles results in 107.5 million tons of carbon dioxide emissions every year.[3]

But growing all of that cotton—nearly 9 billion pounds are projected to be harvested in the US this year, and around 57 billion pounds globally[4]—is done in a particularly impactful way. One major way is in the use of chemicals. About 16 percent of all insecticides are used on cotton.[5] In the last eighty years, cotton yields have tripled because of those chemicals and others—fungicides and fertilizers—and other developments in industrial farming, like better irrigation, increased mechanization, and genetically modified seeds (about half of which are a type known as Bt cotton, which protects against bollworm, although there is debate about its effectiveness and the acquired resistance of bollworms over time to the protein produced because of the cotton's DNA, which is responsible for the protection).[6] While there are some problems with genetically modified or engineered crops (among other issues, it encourages the dominance of multinational corporations at the expense of independent farmers), Bt cotton and new strains being developed (drought-tolerant cotton and cotton that uses nitrogen more efficiently) can make farming more efficient and less reliant on fertilizers and other chemicals, which is what people who are anti-GMO say that they care about. I think it's a real luxury to stand in opposition to genetically modified seeds and, in some ways, scientifically blissfully ignorant. If you disagree, that's fine. Try satisfying your denim habit without GM cotton. I'll wait. In the meantime, let me also point out that cotton production in the US grew by 35 percent from 1980 to 2015, and that globally, pesticide and insecticide use has fallen since the 1990s.[7] I don't mean to say that current pesticide and insecticide levels are acceptable, though. They can and do have significant health effects for the people who are exposed to them, and those health effects are often hidden and unaccounted for costs of agriculture as a whole. And of course there is organic cotton, which is about 0.33 percent of all cotton grown[8] and shares some of the drawbacks of organically grown anything. Some companies also tout their use of "Better Cotton" (which is an initiative to grow cotton "better" with a broad mission of improving labor, land, and water practices[9]) and Fairtrade cotton (which has more of an emphasis on labor practices), but, as implemented, these don't have proven environmental benefits.

All of that sounds bad, I know, but it turns out that the biggest problem with growing cotton is how much water it uses. About 1 percent of global freshwater is accessible (the rest is ice), and about 70 percent of that 1 percent is used for agriculture. Cotton production manages to use 3 percent of the agricultural water supply.[10] Producing one kilogram of cotton requires somewhere between 1,800 and 7,660 gallons of water, depending on where it's grown.[11] (In Europe, it's about 2,610 gallons per kilogram, on average.[12])

In the US, farmers have managed to reduce their water consumption by 80 percent in the last twenty years, which only happened after using so much water that they put the supply of the high plains aquifers, which provide drinking water as well as irrigation, at risk.[13] That's what's happening now around the world; the billions of gallons of water used (and often used inefficiently or outright wasted) to grow cotton is staggering. Not to mention that many of the places where cotton thrives are places whose water supplies are already stressed.

Take the Aral Sea, for instance. Before the middle of the twentieth century, it was the fourth-largest inland water body in the world, fed by two major rivers rushing down from the mountains of Central Asia. In the late 1950s, Nikita Khrushchev, the leader of the Soviet Union, ordered the land around the Aral Sea to be converted to agricultural land, mainly for cotton. In order to do this, they had to divert most of the water from those two rivers away from the Aral Sea.[14] They were incredibly successful: from 1960 to 1980, cotton production there almost doubled,[15] and the Soviet Union produced about a quarter of the world's cotton, most of it from Uzbekistan and Kazakhstan, which surround the Aral Sea.[16] In the 1960s, the sea began to shrink (since there wasn't much water flowing into it anymore); by 1987, it had shrunk so much that it split into two, the North Aral Sea and the South Aral Sea; by 2005, it had lost half of its surface area and three-quarters of its volume.[17] As less water flowed into the area, the quality of the water in the sea(s) declined, the soil in the area got saltier, and the dryness created dust storms, whirling torrents laden with pesticides and fertilizers.[18] Because the soil has become so salty, more freshwater is needed to grow crops, along with more fertilizer and pesticide use. In 1995, the World Bank and the Kazakh government started building a dam to prevent water from the North Aral Sea spilling over into the South Aral Sea, and improvements to the irrigation system along the Syr Darya River were made, too. By 2005, the sea's surface area had expanded by eight hundred square kilometers; the fishing industry has started to come back. Meanwhile, the Uzbek side of the sea is still shrinking, as the government there, which controls the cotton industry, is unwilling to take any steps that might affect cotton production—Uzbekistan is still the fifth biggest exporter of cotton.[19] That seems like a mistake, but don't worry, we've got our best minds on the case: the eastern basin of the South Aral Sea completely dried up in 2015, and Russian and Korean companies have begun scouting for oil in the dried-up seabed.[20]

Not all cotton for denim comes from the Aral Sea region, and that's a particularly harrowing example of agricultural intervention gone awry, but I just wanted you to know, because otherwise

I have to be alone with that information. Also, since just over one-third of all the cotton produced in the world is used to make denim,[21] there's a chance that some of the cotton in your jeans (or something else you own) could come from Uzbekistan. All of that cotton is spun into yarn and woven together into billions of meters of textile. (Some jeans, particularly in the last few years, as jeans—especially women's jeans—have become skinnier and stretchier, also contain synthetic fibers, like elastane.)

Beyond all of the water used to grow the cotton, producing the denim fabric requires even more water: as much as 2,900 gallons can be used to produce a single pair of pants (using conventional methods), mostly because of the dyeing and finishing.[22] Yarn can be dyed in as many as twelve dye baths to turn it your standard denim blue.[23] If the water used in the dyeing process is released to the environment, it can cause water pollution as well.

Then there are a variety of other finishing treatments in the dyeing stage, all of which might require washing. In the 1970s and 1980s, hip fashion people decided that denim shouldn't look like denim anymore, and they came up with stone washing and acid washing. I didn't realize this, but stone washing sometimes literally means that jeans are put into washing machines with stones (pumice), adding to the environmental footprint of denim: the stones have to be mined from volcanic ash and shipped to the factory, plus they damage the machines and make the fabric less durable. Some companies use a chemical substance to simulate the effect of rocks. To acid wash a pair of pants, you soak that pumice in bleach before you wash it, also a good and smart idea.[24]

Most of the water used to dye or finish denim is not recycled, so it ends up in the environment, where it can pollute drinking water supplies or harm aquatic animals. If textile factory workers are exposed to some of the chemicals—which can happen if there are inadequate safety measures in place, like a lack of proper ventilation—that can cause health problems as well.[25] In Xintang, the denim capital of China, rivers have been stained dark blue by water gushing out of factories, which villagers have reported causes their skin to itch and fester, according to Greenpeace.[26] As global water demand increases, it will become even more important for the agriculture and fashion industries to figure out how to use water better. If we don't change anything about the way we're using water, the global demand for it could exceed our supply by about 40 percent by 2030.[27] Currently, humans are rapidly consuming groundwater without knowing when it might run out, especially in some of earth's driest places. According to an assessment of data from NASA's GRACE mission, between 2003 and 2013, more than one-third of our largest aquifers were depleted without enough water coming in to replace what was taken out.[28]

Recycling denim can be tough, too, because of the stylistic things we've added to our jeans: copper rivets and zippers, which can be difficult to remove in the recycling process or are not recovered, create a large amount of metal waste worth around $110 million each year.[29] When you read later on about cargo shipping, you'll see that there is also a big cost in getting the cotton from (for example) Texas, where it's grown, to Indonesia to be spun into fibers, to Bangladesh to be made into denim, and sent back to the US to be sold.

There are ways to make jeans using less water, and a lot of companies have recognized this and been working on it. Levi's started the Water<Less program, which does exactly what you'd think it does, and has helped save, by their accounting, more than 525 million gallons of water and recycled another 52 million gallons.[30] In 2017, Levi's made more than half of their jeans with the Water<Less techniques,[31] which the company has shared with others in the industry. Their plan is to make 80 percent of their

products using the Water<Less system by 2020. For some styles, they've managed to use 96 percent less water.[32] I also tried to speak with Re/Done, another company trying to limit its resource waste, in this case by taking vintage Levi's apart and remaking them, but they stopped responding to my emails.

I did, however, correspond with Everlane, whose representatives eventually did answer my emails. They produce their denim at a factory in Vietnam—okay, not great on the carbon footprint/transportation thing—but they are trying to produce denim in a responsible way. The factory, Saitex International Dong Nai Co. Ltd., recycles 98 percent of its water to a "drinkable" state, air-dries the denim (instead of using a dryer, which uses a lot of energy to heat the air enough to dry the jeans), and turns the excess denim it produces into bricks for affordable housing.[33] The factory loses only 0.4 liters of water for each pair of jeans produced versus 1,500 liters in conventional production, and uses rainwater (instead of river or lake water) as its source. Renewable energy supplies at least some of their electricity, through solar panels and miniature wind turbines. Everlane's denim is also sold at a reasonable price point: $68, because they are satisfied with marking up their products by two or three times, rather than the industry standard five or six times.*

According to Everlane's email, they audit the factories that produce their clothes four times a year, and their production team also visits the factories a few times a year. If the factories don't score at least 85 percent on their audit, they work with them on a plan to improve, and if they continue to fall below the standard, they move to a different factory. Everlane's denim production was

---

* I also wanted to talk to Reformation, which talks a big game about its sustainability—they note the amount of water saved and emissions avoided through things like vintage shopping, buying textile discards from distributors, etc.—but they told me they don't "participate in individual projects," which, obviously, is their giant loss.

their first "true entry" into sustainability, they told me, focused on reducing water and energy use. I'm not sure, then, about the sustainability of their other clothing, which gets poor ratings from some sustainability groups, but they certainly advertise their efforts well. However, the company's philosophy, at least as expressed to me, is moving in the right direction: "We know customers care, but they don't fully understand the implications of production. Companies need to lead the change and educate the world on the issues and how they can be improved. Any company that is not doing this is actively choosing not to improve our environment."[34]

Unfortunately, a big part of the environmental impact of clothing comes from how we wear and use them, too, which I'll get to later on in this section. The main piece of that is washing our clothes, and we could stand to wash our jeans less, like not after each wear. If we washed our jeans after every fifth wear, according to Levi's, "it would move the needle."[35] The company also claims that hanging them out in the sun if they aren't particularly dirty also works,[36] which, okay, but I live in New York. I'll do what I can.

Athleisure Forever!

Imagine a pair of yoga pants, or a fleece. (You probably don't have to imagine these things—you probably are wearing them right now or at least own them, given the current state of how we all dress now, which is, a lot of the time, in our gym clothes.) While you picture these things, or feel their stretchy smoothness or general swaddling-softness on your skin, try to imagine what they're made of.

It's oil. They're made of oil.

Whether it's polyester, fleece, spandex, elastane, nylon, or acrylic, our clothing is made, more and more, of crude oil that is turned into polymers (chains of identical molecules linked together), which are then whirled into fibers and knitted together to make fabric and sewn into the shapes that we wear all the time. More than 60 percent of all of our textile fibers are now man-made synthetics, derived from oil, a dramatic change over the last few decades.[1] It was only around 2007 that synthetic fibers, mostly polyester, surpassed cotton to make up the majority of the fabrics we wear.

You could talk about almost any type of clothing and find a way to talk about synthetic fibers, but I want to talk about athleisure, because it's a section of the fashion industry that meets us where we are—comfortable, looking like we want to be active but not really that active, since if we were actually training for even a 5K race, we would probably be wearing actual technical apparel (also made of synthetic fibers) and not a pair of leggings that look enough like clothes to wear around while we run errands—and is, perhaps not coincidentally, an industry that has ballooned during the short period of time since synthetics became the dominant textile fiber.

Compared to cotton, synthetic fibers require a lot less water to produce, but that's not necessarily a good enough argument for using them, since they have other significant impacts: they are still made of oil, and their production can require a lot of energy. MIT calculated that the global impact of producing polyester alone was somewhere between 706 million metric tons of carbon dioxide, or about what 185 coal-fired power plants emit in a year.[2] Samit Chevli, the principal investigator for biomaterials at DuPont, the giant chemical company, has said that it will be hundreds of years before regular polyester degrades.[3] Plus, while the chemicals used in production typically aren't released to the environment, if factories don't have treatment systems in the last phase of production, they can release antimony, an element that can be harmful to human health, as well as other toxins and heavy metals.

Despite having just written a good amount about the impacts associated with the production of synthetic fibers, that's actually not why I wanted to call attention to your yoga pants and dry-fit sweat-wicking T-shirts, which we wear out to dinner. It is hard for me to leave my fashion critique at the door, but what I actually want to say about synthetic fibers is that they are everywhere—not just in all of our clothes, but literally everywhere: rivers, lakes, oceans, agricultural fields, mountaintops, glaciers. Everywhere. Synthetic fibers, actually, may be one of the most abundant, widespread, and stubborn forms of pollution that we have inadvertently created.

Austin Baldwin, a hydrologist with the US Geological Survey, pulled some quagga mussels, Asian clams, smallmouth bass, and carp out of Lake Mead, the lake created by the Hoover Dam to bring water to people in Arizona, California, and Nevada, southeast of Las Vegas.[4] Though I'm sure those things could be prepared in a way that is delicious, that's not why he was gathering them from the water. He was doing that to *preserve* the species, storing up a vital source of protein for the coming apocalypse, when the world inevitably explodes, and then the explosion collapses in on itself, imploding and sucking the universe into a somehow-fiery black hole. Just kidding, although who knows. He was actually cutting out the guts of the fish and examining the whole shellfish organism and looking for one thing: microplastic fibers. In the shellfish, he found somewhere between one and ninety-eight plastic fibers in each organism. In the fish guts, there was one fish that had no fibers. The others had somewhere between one and nineteen plastic fibers. The fibers mostly ranged from clear to blue to black to red, and Baldwin said that in the muscle bodies of the shellfish, it was possible to see the fibers without looking under a microscope. In the fish, it was more difficult, because the fibers had likely woven themselves into the intestines of the creatures, insinuating themselves into the very *fiber* of the fish's being. (The fibers were only examined after they were separated from the fish.) None of this was surprising to Baldwin.

But it was surprising to me: What? Plastic? In fish? And in fish with shells? And also in a lake? How did the plastic get there? It's not that Baldwin was expecting to find all of that plastic based on a hunch. It's because he'd seen this before. About fifteen years ago, scientists started noticing that microplastic fibers were showing up on shorelines around the world and in samples from the surface of the ocean. Many of us may have heard, because of viral photos and much of the coverage around the straw ban and the Great Pacific Garbage Patch, that ocean plastic is a problem. What many people don't appreciate is that most of the plastic in the ocean isn't a whole straw, or even a whole plastic bottle. Much of the plastic in the ocean is tiny pieces, known as microplastics (often defined as less than five millimeters long, sometimes less than one millimeter), mostly from fishing equipment. Besides that, ocean plastic, for the most part, has been broken up by ultraviolet radiation, wind and waves, tide and time. Some of them were micro before they got to the ocean, namely microfibers (and microbeads, too). As it happens, about 85 percent of the plastic pollution found on shorelines around the world is in the form of microfibers.[5]

Given that scary statistic, scientists were also starting to think that if plastic was getting to the ocean, it might also be getting into freshwater. Among the first scientists to test that theory was Sherri Mason, a professor at the State University of New York, Fredonia, who took a boat out onto the Great Lakes to sample for microplastic fibers. She found, in the water of Lake Ontario, levels of microplastic pollution as high as in the most polluted areas of the oceans; in one square mile near Toronto, she found 3.4 million microplastic particles. (In her study, she mostly found microbeads, which were all the rage around that time and have since been banned from our cosmetic products.)[6] She didn't find a lot of microplastic fibers, which she said was surprising. Soon after Mason's study was published, Baldwin received a grant from the Great Lakes Restoration Initiative at the EPA to study pollution in the tributaries of the Great Lakes; he asked Mason if she would help analyze their samples for microplastic pollution, and she agreed. During that experiment, Baldwin said they were all surprised by the results from these tributaries: fibers were everywhere![7] (After another study, Baldwin wrote a paper for the National Park Service titled "Microplastics Are Everywhere!") Studies from rivers all over the world started showing the same thing. Baldwin said that he and other scientists are trying to figure out why they are finding fibers more often in rivers than in lakes.

They aren't completely sure yet, but he had a hypothesis: microplastic fibers often sink, but the churning in a river might keep them from sinking to the bottom, so you might get some fibers from sampling the surface. In lakes, there are higher concentrations of microplastic fibers in the sediment, which lends some credence to this theory.

The inevitable next question is, of course, where are the microfibers coming from? They might be coming, as you perhaps guessed from the beginning of this section, from our clothes. (And before you ask, yes, scientists know that clothing is an important source of microplastic fiber; it's not just other types of plastic, already in the water, breaking apart into smaller pieces.)

When you put your clothes in the washing machine, tiny fibers—polyester, acrylic, nylon, spandex—whirl off into the water, swirling around until the water is sucked out through the filter. You probably don't notice this, at least not for a while, because the pieces that break off are so small that you can't see a difference in the clothes you've washed, though over time you may notice that your clothes get thinner, and the microfibers don't seem to appear inside your washing machine, either. Rest assured, your clothing is shedding fibers. (Lint in dryers is also made up of clothing fiber, but your dryer isn't connected to a water supply.) A study by researchers at Plymouth University from November 2016 found that more than 700,000 fibers were released during an approximately thirteen pound load of laundry.[8] (Natural fibers, like from cotton or wool, were found, too, but synthetic fibers shed the most, particularly polyester, and fleece most of all. A fleece garment, this study found, could release as many as 100,000 fibers in a wash; a garment made of another kind of fabric might release about 900 fibers.) Another study, performed at the University of California Santa Barbara in partnership with Patagonia, found that a city with 100,000 people might release somewhere between 9 and

110 kilograms of microfibers per day into local rivers, canals, and streams (an amount which adds up to as many as 15,000 plastic bags).[9] There are some other characteristics that might affect how much a particular garment sheds: the age of the garment, the type of detergent used, the temperature of the water, the type of washing machine (front-loading machines cause less shedding than top-loading ones because of the placement of the agitator, the real name of this part of the washing machine), the original quality of the product (better-made things shed less, but fleece made from recycled plastic sheds less than fleece from freshly made polymers), and how many times the garment has been washed (clothing will shed more at the beginning than later on).[10]

These microplastic fibers are small enough that they mostly pass through the filter and travel through pipes and, if you live in the developed world, into a sewage tank or a wastewater treatment plant somewhere nearby-ish. In a wastewater treatment plant, the vast majority of fibers (somewhere between 75 and 99 percent) settle into the sludge, which I agree is gross, but that's what happens. Sludge from wastewater treatment plants can be used as agricultural fertilizer, and it is slathered (a cool word that makes you want to eat food) onto fields. From here, the fibers can get into the soil and, from there, into the groundwater or into the food that our food eats. But more directly, for our purposes, the fibers can run off the fields (as we talked about with fertilizer) and make their way into waterways, such as the rivers, streams, and lakes that form the tributaries to the Great Lakes (or elsewhere), where Baldwin and Mason found them.

Even more directly, fibers that don't settle in the sludge may leave the treatment plant in the wastewater, depending on how small they are and how fine the filter is. For the most part, they don't get out—Mason has estimated that less than one piece of plastic might escape from the plant with every gallon filtered.

Even so, a lot of water passes through a treatment plant, so about 4 million pieces of plastic can escape every day, 60 percent of which are microfibers.[11] In Baldwin and Mason's study of Great Lakes tributaries, there was not a positive correlation between the presence of wastewater treatment plants and the presence (or not) of plastic fibers,[12] but this doesn't necessarily mean that wastewater treatment plants don't release any plastic fibers to the environment. (The original focus was on the ocean because a lot of plastic fibers were going into the ocean, and because many wastewater treatment plants feed into the ocean and not into freshwater. A 2017 report found that at least 35 percent of the microplastic fibers that enter the ocean come from synthetic textiles, and that likely doesn't include the fibers that end up in the deep ocean or on the ocean floor.)[13]

Some microfibers may also come from the air, believe it or not. They may get whisked off the clothes we're wearing as we move through the world, swirled up in the air, and deposited somewhere else: land, lakes, rivers, oceans. A small study of the River Seine basin, which includes Paris, found that 29 to 280 microplastic particles fell from the sky per square meter each day, which could add up to as many as 327 billion particles every year in that region.[14] How you say, ooh-là-là . . .

And aside from clothing, research from Europe and North America suggests that tens of thousands or hundreds of thousands of tons of microfibers are added to farmland through sources other than fertilizer, like plastic sheets laid over soil to keep moisture in and prevent weeds from growing. One scientist described soil in Australia as "glistening" from mixed waste plastic. All of this plastic could affect soil health, changing how it retains moisture over time or facilitates pesticide concentration and, like the plastic from the fibers, gets into the groundwater, etc., etc., etc.—we get it.[15]

The reason I'm writing about athleisure is, in part, because athleisure clothing is made almost exclusively from these synthetics. (It also sometimes includes nanosilver fibers, which seems, I don't know, wasteful, but these fibers have also been found in the environment and can cause harm to aquatic organisms.) Athleisure ostensibly is designed for (or at least marketed to) people who care about the outdoors or being outside or at least improving their bodies, which can only truly happen in a world with clean air and water, a world where all bodies are temples, hand-sculpted from the finest marble. For instance, Patagonia's fleeces are made the way they are in order to answer another question which was, what do we do with plastic bottles? So they tried to make fleece from plastic bottles, and they did, which seemed great.[16] Now, that plastic is getting back into the environment in the form of microplastic fibers, much to the chagrin of Patagonia (which is funding studies, like the one from UCSB, and offering solutions to their customers while they try to figure out how to get rid of microfiber plastic pollution in the meantime).[17] Less invisibly, the demand for used plastic bottles, from Patagonia and others, is also beginning to outpace the supply. So some manufacturers, according to some reports, are buying new bottles to make polyester textile fibers that can be called "recycled." Which is nuts.[18]

We don't know the long-term effects of plastic in the environment; it hasn't been around long enough for us to have a full understanding of what will happen over the decades- and centuries-long lifetime of the plastic we've made over the last seventy years (and the plastic we continue to produce). We already know that it is everywhere: microfibers and other forms of microplastic are in Arctic sea ice; plastic has been found on top of a glacier in North Cascades National Park in Washington State, on an uninhabited island in the South Pacific, in the Mariana Trench (the deepest part of the ocean), and in Antarctica.[19] A lot of the plastic is, of course, not from our clothing. But some of it is.

What does this say about us? To me, it is, on the one hand, a testament to the amazing power and unconscious voracity of human civilization, settlement, and creation. That plastic, which has not been around that long, is in all of those places is astounding in a horrible way. I don't fault anyone for not anticipating, when we started weaving plastics together to make clothing, that plastic would then, irrevocably, spread to every corner of the earth. What I do think is a problem is how much of it there is, which is partly our fault, because we want more and more things, including but not limited to athleisure. But now we know it's a problem, and the companies that make these things are not doing enough, if anything, about it. The proliferation of stuff and our role in buying, using (or not), and throwing away that stuff is truly an enormous part of the problem of pollution in general and plastic pollution specifically, and I look forward to bumming you out with it in the next chapter.

Fast Fashion, but Going Nowhere

Consider, if you will, the fashion industry. Take the cotton fibers woven together into a tight denim sausage casing for your leg (thank you to whoever decided we should all wear skinny jeans), and the synthetic fibers gently floating off your back into the air you breathe and the water you drink. Imagine that each of these fibers, millions of them just in the outfit you're wearing today, is a paint stroke. Imagine using all those paint strokes, each one a fiber, together to create a painting. That painting, that masterwork, that Sistine Chapel fresco of fibers, is fast fashion, but just the part that's in *The Last Judgment*—the extra hellish parts—because that's what your descendants will think of this ever-bulging sector of the fashion industry—that it certainly portends a hellscape, if we're not living there already.

That beautiful and poetic prelude was a way to say that without cheap textile production fast fashion wouldn't be possible. And fast fashion, as it happens, presents a particularly environmentally damaging sector of the global economy, largely because of the scale.

What do I mean by fast fashion? I mean a brand, usually with an actual retail presence (physical stores) as well as some online presence, that sells clothing whose designs are largely derived from high(er) fashion or luxury brands but sold for much less money. They produce lots of clothes all year round and are able to produce clothes quickly, so what appears on the fashion runways can also appear relatively quickly on their racks. I think you all know the brands I mean: Zara, H&M, and nearly all of the brands they own, ASOS, Uniqlo, Topshop, Forever 21, etc., etc., and on and on forever.

Though some of these brands have existed for decades (H&M started in the 1940s),[1] the phenomenon of fast fashion as I'm talking about it has only really existed since the beginning of the twenty-first century, but it has exploded: Inditex (which owns Zara) and H&M are currently two of the largest global clothing retailers.[2] H&M has been struggling for the last few years, ending 2017 with $4.3 billion in unsold clothing and having to shut some of its stores. This is not, however, the end of fast fashion. Instead, some argue H&M is no longer fast enough to keep up with some other brands, especially those, like ASOS, that are online only.[3]

How did we get to this point? Part of it has to do with synthetic fibers and the relative ease and low cost of producing them. Synthetic fibers, unlike cotton, don't need to be grown and harvested at the end of a growing season. Part of it also has to do with China's willingness to open itself up to the global economy and subsequent explosion of the garment industry there. There are 15,000 textile factories in China alone, but as China begins to crack down (a little) on the environmental and labor practices of some of these mills, they've moved elsewhere: Vietnam, Bangladesh, Cambodia.[4]

Both of those things combine to make it possible for clothing to have become cheaper relative to other consumer goods. And the quick turnaround in production has made it possible for some of these brands to outpace the natural cycle of the seasons. You may recall that there are four seasons: summer, fall, winter, and spring. In fashion, there used to be two: spring/summer and fall/winter. Now, because clothing can be mass-produced within weeks of being imagined, fast fashion chains refresh their stocks constantly. Zara has between sixteen and twenty-four seasons a year; H&M has something like eleven, although new products ship to stores almost daily. And fast fashion chains have changed the retail landscape more broadly: the average European fashion line now has five different seasons a year, which have almost nothing to do with actual seasons and are just excuses to push out new clothing to get us to buy it.[5]

So there are more different things and we can buy them for less. And we buy all of them. Compared to 2000, we bought 60 percent more clothing in 2014. That year, according to a McKinsey report, the fashion industry as a whole produced more than 100 billion garments, enough for every person on earth to have about thirteen new pieces of clothing each year.[6]

The constant consumption of their clothes is keeping these companies afloat. The clothes stay cheap enough that we think we can afford them, because, obviously, all of the external costs of making clothing—the environmental pollution, the health impacts for workers and for those who live around factories—aren't paid for by the companies who make the clothes or the people who buy them: us. The World Resources Institute calls the problem of unchecked consumption the "elephant in the boardroom," writing that trends like fast fashion stoke consumer expectations for more stuff, while the companies pay an "alarming lack of attention" to the limits of our natural resources, many of which are essential to the production of clothing and other *stuff*. They write that we are already "close to the limits of the planet's ability to provide," and that business-as-usual would mean three times as much consumption of our natural resources as already occurs, which is way too much.[7] If 80 percent of people in the developing world shopped like people in the US and Europe, we would see a 77 percent increase in carbon dioxide emissions associated with clothing production, a 20 percent increase in water usage, and a 7 percent increase in land use. It's already starting to happen: according to that McKinsey report, in five large developing economies—Brazil, China, India, Mexico, and Russia, clothing sales are growing eight times faster than in Canada, Germany, the United States, and the United Kingdom.[8]

And what do we do with that excess of stuff we now have? Do we treasure it and thank our lucky stars that we can buy an imitation Gucci bomber jacket for $10 and kiss the ground and love our parents and their parents for putting us on this verdant, splendid earth? Yep. That's how the fast fashion industry thrives and survives: gratitude.

Nope! That's not how. We throw it away. When we're buying fast fashion (which we don't have to do), we actually have to buy more, because the clothes aren't made well. They're made cheaply and quickly, so they don't last as long. We get rid of about 60 percent of the clothing we buy within a year of its being made;[9] we used to keep our clothing at least twice as long. In the United States, we generated sixteen million tons of textile waste in 2015, up from eleven million just ten years earlier.* Only about 14 percent of clothing and footwear is recycled.[10]

Clearly, that represents a staggering waste of resources and energy, and it means that the people who are exposed to the pollution associated with producing the raw materials for clothing or with combining those raw materials to make the piece of clothing you wear a few times and then throw away are suffering for basically nothing, never mind that they shouldn't be suffering at all. You may think, *Well, I donate my clothes,* or *I heard about a program that takes jeans and makes them into insulation,* or *What about recycled textiles or all of the clothes that we send overseas in the form of aid?*

---

* While the main source of textile waste is from discarded clothing, the EPA also includes rubber and leather in textile waste, as well as furniture and carpets.

All of those things happen, but not to the extent that you think, and sometimes with surprisingly negative consequences. At this point, though, is it really a surprise?

According to the EPA, we donate or recycle only about 15 percent of our clothing in the US; the remaining 85 percent ends up in a landfill or being incinerated, composing just over 5 percent of our municipal solid waste.[11] In Japan, it's 12 percent, and in China, just 10 percent. Some other countries do a better job: in Germany, about half of all discarded clothing is reused and a quarter is recycled;[12] in the United Kingdom, there have been efforts to reduce the lifetime impacts of clothing (washing, drying, and ironing less or at lower temperatures) and clothing waste, cutting it by 50,000 metric tons between 2012 and 2017 (though the carbon footprint of clothing grew overall because people are buying more clothing).[13]

Some companies, like H&M, will advertise that customers can bring back their clothing—any clothing, not just H&M clothing!—to an H&M location—any location!—and the company will recycle them, and that customer will get a voucher to buy more H&M clothing—any H&M clothing! In 2016, H&M collected 1,000 tons of clothing for recycling. Since recycled yarn makes up only a small percentage of an average piece of new clothing made from recycled material, it could take H&M more than a decade to fully use up 1,000 tons of fabric waste, according to Lucy Siegle, a columnist for the *Guardian*, not to mention that H&M produces tens of thousands of tons of clothing (if not more) a year.[14] It's not that it's necessarily bad, but it's somewhat misleading, since the campaign insinuates that all of the clothing will be recycled to make new clothing. Meanwhile H&M is encouraging people to buy even more for less.

Recycling technology for clothing is not great. If a piece of fabric is made of two different types of fibers—cotton and polyester, let's say—it's pretty difficult to separate them. If fabrics are recycled, they are usually mechanically recycled, which degrades the quality of the fibers (they are chopped up into shorter pieces, so they lose strength), and generally "downcycled," meaning they're made into things other than clothing, like insulation, rags, or carpet padding. Textiles that are chemically recycled can be recycled more or less infinitely, but that process is much more expensive than mechanical recycling, so it is not viable on a large scale and uses more energy. Plus, there is the risk of chemical pollution, too.[15]

If clothing made completely or partly from synthetic fibers goes to a landfill, it doesn't decompose, since it's made of plastic. Cotton (and other natural fibers) will eventually decompose in a landfill, but they'll release methane, a powerful greenhouse gas, and carbon dioxide.

Donating clothes can have unexpected consequences, too. Even though, in the US, about half of donated clothing (which is not that much, remember) is reused here, a lot of it is also exported, generally to the developing world. According to Andrew Brooks, a lecturer in development geography at King's College London, used clothing sent to the developing world can depress local textile and clothing production and markets, flooding them instead with (often extra-large) clothing and creating a further dependency between rich countries and poor ones.[16] Also, we do stupid things like make championship T-shirts (and lots of other clothes) for both teams in the Super Bowl or other sports finals, and then ship the losing teams' shirts abroad. Those clothes are created with the express purpose of wasting them—excuse me, donating them to people in need—needlessly using water, land, oil, chemicals, energy, and labor, all so you could buy your "[Insert Team Name Here] are #1" as soon as the game is over, just in case.

But, you now exclaim, wildly and loudly into the void (since I can't hear you), "What about closed-loop production?" momentarily thrilled to have noticed a flaw in my genius master plan to bum you out. Meanwhile, I know all about closed-loop production (sometimes called the circular economy). Or I know a little bit about closed-loop production: it's basically the idea that you only make clothing (in this case, but the principle applies to everything) from reused materials and keep reusing them forever or, if they are natural materials, turn them into biodegradable waste. It also requires recycling water and chemicals used in the manufacturing process, so that you keep using them forever, or for as long as you can. So far, it doesn't really happen, although it could be a good thing. (Certain types of fibers, like lyocell/Tencel, coming right up in the next chapter, are made in a closed-loop system, but primarily to recycle the chemicals used in processing.) Unfortunately, even if closed-loop production were to actually happen, it doesn't get you very far in terms of saving the world. First of all, it's expensive, at least in the beginning, and it doesn't result in anywhere near as much energy savings as does using renewable energy or making efficiency improvements. For instance, according to a report from the sustainability consulting firm Quantis, using 40 percent recycled fiber yields only a 6 percent decrease in climate impacts, a 4.4 percent decrease in water consumption, and 3 percent decrease in human health impacts. By contrast, if a company were to get 78 percent of its energy from renewable sources or improve its efficiency by 72 percent, the emissions of clothing manufacturing would be reduced by 50 percent.[17] That scale of reduction just isn't possible with closed-loop manufacturing. Linda Greer, our clothing expert from NRDC, says she thinks the circular economy (or closed-loop production) is basically the industry's "fake way" of addressing fast fashion—the idea that "if it goes around and around it's not that bad"—while still continuing to make a lot of clothes that we, for the most part, don't need.

What would be more effective, Greer argues, is if people had their core wardrobe of the things they really needed that would last, and then the rest we would share, a model used by Rent the Runway, for example, which allows customers to pay to have clothes for a certain amount of time and then send them back so they can be washed and worn again by someone else. That, perhaps, may be a more effective model of going "around and around."

It can be hard to argue in favor of more expensive clothing, even if it lasts longer, because it risks excluding some people with less disposable income from buying clothing, which has been considered a basic human need since Adam and Eve ate the apple and fig leaves became the bare minimum. That's valid, and the environmental movement and efforts toward less consumption won't succeed on the scale required if we don't make room for everyone. But I also think it's important to remember that fast fashion clothing is not made to clothe people without means—it's made to get people of all income levels to think that they can afford buying lots of clothing that will not last, meanwhile extracting lots of finite resources that no one can afford to waste. We all need to focus on making and buying clothing that doesn't exact such a heavy toll on the environment—our clothes will be less expensive and last longer, and we will have to pay less, in the end, for the ultimate costs of pollution, climate change, and disease. I can almost guarantee you don't need that acrylic floral onesie. Yes, I know it looks cute, but does it look *that* cute?

It's Not Wood, It's Rayon

If you hear the words *low-viscosity rayon* and you are like me, the first thing you think of is *Legally Blonde*, specifically the indelible, Oscar-worthy scene in which we all learned and never forgot that you cannot use half-loop top stitching on the hem of low-viscosity rayon. It will snag the fabric.

However, there is more to know, hard as that might be to believe. Things like, what is rayon, and where does it come from? Given what this book is about and the fact that rayon makes an appearance, is it bad for the environment? Most importantly, was Elle Woods right?

Rayon (which, almost all the time, is viscose rayon; it's unclear if low-viscosity rayon is actually a thing) is one of the first man-made fibers, and it is made from plants and trees—specifically from cellulose, a natural polymer. So far, it sounds eco as hell. Early forms were developed in the nineteenth century as a substitute for silk[1] but now it is the third-most-used textile fiber,[2] after polyester and cotton, making up about 7 percent of all of our clothing fiber. It's a relatively small amount, but it's growing: its share of fabric textiles is projected to reach about 8.5 percent by 2030.[3] Again, relatively small, but if there is a stone on our path to knowledge about the fashion industry, I'm not going to leave it unturned.

Today, rayon is mostly made from wood, mashed up and dissolved into a pulp, then spun and stretched out into the fibers that make up your clothing. It can come from different kinds of wood from different kinds of forests all over the world, but it is often processed in textile mills in China or India, though mills are popping up in other parts of Asia as well. Producing rayon uses a lot of chemicals, including carbon disulfide, which can cause serious health problems to workers who have chronic exposure to the substance through inhalation, including neurological damage and reproductive effects.[4] These chemicals can also be released into the environment, though it's not entirely clear what the environmental effects or health effects from environmental exposure are. In the textile industry in general, while the Chinese government is beginning to enforce environmental regulations and punish bad actors, there is still a lot of environmental pollution, and a lot that comes from rayon production in particular. And if the pollution isn't happening in China, it's likely happening in Vietnam, Indonesia or other countries where factories move when China (China!) gets too strict and the costs of production rise.[5]

As you'll see when we talk about biomass, if the wood being used to make rayon was left over from other processes—like from lumber for construction—it would be more sustainable. The thing about forestry and logging, though, is that it's mostly not practiced in a particularly sustainable way.

You may have heard about deforestation in Indonesia, where ancient, old-growth, endangered trees in the rainforest—particularly productive, in terms of biodiversity and the number of species they support, and able to sequester a lot of carbon dioxide—are cut down or burned to make way for tree plantations—sometimes bamboo, sometimes palm trees. If you have heard about it, it was probably in the context of palm trees for palm oil.[6] According to a report from SCS Global Services, an environmental consulting and research firm, forests in Indonesia went from being undisturbed twenty years ago to "fully disturbed" today. Over the next two decades, the organization anticipates that there will be "virtually no undis-turbed forest in Indonesia." A lot of the wood from the trees that are destroyed in the process is used to make rayon (though it could be used to make other things, too); the wood from the plantations that replace those ancient rainforests, also not able to sequester carbon dioxide once they are cut down, can also be used to make rayon. In the same report from SCS Global Services, of ten different sourc-ing and production scenarios, logging in Indonesian rainforests and plantations was the worst in terms of its effects on climate change, regional acidification, resource depletion, and human health.[7]

And it doesn't just happen in Indonesia: according to Canopy, an environmental group that tries to get forestry companies and those they supply to protect the world's forests, about 30 percent of rayon produced for clothing is made with wood pulp devel-oped from endangered and ancient forests;[8] the Rainforest Action Network has found that about 120 million trees are cut down to make our clothes every year.[9] Some examples: pulp from Cana-dian boreal forests is also an important source of rayon, and comes second only to logging of the Indonesian rainforests as having the worst impacts on global climate change.[10] Logging of native forests or destruction of grasslands in South Africa to create tree plantations (as well as in Indonesian rainforests and plantations) causes significant disturbances, too: the plantations become crowded with trees that are not necessarily native to the region or are grown in a way that the ecosystem would not otherwise support. However, these forests are more productive than those in Europe or Canada (other areas from which pulp for rayon is sourced), so they require less harvesting to produce the same amount of pulp.[11]

Plus, rayon, because it's made of plants, and specifically rayon made from bamboo, is often fodder for greenwashing (when com-panies market themselves as more environmentally friendly than they actually are), since the whole thing sounds, at first silken glance, eco as hell. But the Federal Trade Commission didn't buy it! In 2009, the FTC settled claims with four companies who labeled and marketed their products as having been made with bamboo fiber when it was really rayon and promoted "green" claims that the textiles were manufactured in an environmentally friendly way, that they retained the "natural antimicrobial properties" of bamboo, and they were biodegradable.[12] The agency also alerted manufacturers and sellers not to claim that their textiles were made from bamboo if they weren't actually made from woven bamboo fiber. The following year, they sent a warning to seventy-eight companies, including Amazon, Macy's, Sears, Target, and Walmart to stop labeling rayon products as bamboo.[13] Amazon claimed that products like "Summer Infant Crib Sheet" were "100% Organic Bamboo."[14] This was, strictly speaking, not true. Amazon, Leon Max, Macy's, and Sears ignored the warning, despite having been told they could face civil penalties. In 2013, those four companies settled with the government, paying a combined total of $1.26 million.[15]

There are better ways to produce rayon from bamboo. If made mechanically instead of chemically, sometimes known as "bamboo linen," it has a relatively small environmental impact, though it costs much more. Another type of rayon fiber, known as lyocell or Tencel, can be made from bamboo, using a different chemical thought to be less toxic,[16] though studies are scarce.[17] There are some genuinely greener options: rayon made from Belgian flax (which also has a smaller transportation footprint because the flax is grown and milled in Belgium and doesn't need to be pulped), and recycled pulp from clothing pulped in Sweden and milled in Germany.[18]

But that stuff can be hard to find because it's harder and more expensive to produce, and regular rayon is common, and increasingly so. There's an additional problem beyond the deforestation, though, which is the chemicals. I already mentioned carbon disulfide, which can cause insanity, kidney disease, heart attack, and stroke under chronic exposure. Hydrogen disulfide can be produced during the spinning of the fiber and can cause irritation of the eyes as well as neurobehavioral changes. Sodium hydroxide (also known as lye), also used in production, can burn the eyes, skin, and internal tissues if inhaled or ingested. Sulfuric acid, another production chemical, can be harmful when inhaled and if combined in mist form with other acid mists (the dewy, refreshing mix of a corrosive chemical spray) can be carcinogenic.[19]

Though the world's largest rayon manufacturing company, Aditya Birla Group, was rated number one by Canopy for its work on conserving and protecting old-growth and endangered forests in its sourcing efforts—good for them!—a report from the Changing Markets Foundation also found that, in the areas outside its twelve mills, the company was releasing chemicals into the air and water. At the Grasim plant in India, an air sample taken by the report's investigators showed that levels of carbon disulfide were 125 times higher than the World Health Organization's guideline value. Though difficult to tie directly to the chemical pollution, people living near the factories were suffering from cancer, tuberculosis, and reproductive issues; local agriculture had also been destroyed. In West Java, their Indo Bharat rayon plant discharged water so contaminated that it did not even meet Indonesia's "worst-in-class" water quality standards, which means it shouldn't even be used for irrigation. Near the factory, the organization found children playing in the water, which was also used for washing and cooking. Aditya Birla Group denied all claims in the Changing Markets Foundation report, and said that wastewater and air emissions at its Grasim plant met "all applicable norms."[20] Nonetheless, since the investigation, Aditya Birla Group and

Lenzing, an Austrian viscose company that produces about 19 percent of the world's rayon, have committed to using a closed-loop manufacturing system by the early part of the next decade.[21] In order for us to trust that closed-loop production actually does work as promised, companies have to make a lot of substantial changes rather quickly: where the material is sourced from, if the products made from the recycled materials are good quality (otherwise people won't buy them and the market won't sustain itself), etc.

And either way, as Linda Greer said, if it's made by using all of those chemicals, rayon shouldn't count as a natural product. The chemicals, whether or not they get into the environment (which is a big *whether*), can prevent the material from decomposing, during which process they would emit methane anyway, which is why we can't have nice things the end.

The Yarn That Makes a Desert

The Gobi Desert cuts a sandy stripe from the eastern edge of Mongolia to the center of northern China, blanketing much of the border between these two countries in dunes of dust. It's a climate of extremes: it receives a maximum of two inches of rain a year in some parts, eight inches in others, with extreme winter lows reaching to −40°F and summer temperature sometimes exceeding a blistering 110°F.[1] Despite its harshness, it is a delicate ecosystem: there is not a lot of water; plants are scattered; winds bearing down from Siberia rage for most of the year. And the desert is getting bigger—estimates suggest that, between Mongolia and China, the desert is growing by somewhere between 900 and 1,500 square miles every year[2]—partly because of climate change, partly because of goats. Yes, you heard me correctly: goats.

On the grassier plains of this desert at the top of the world, nomadic herders have been shepherding their goats (as well as camels, sheep, horses, yaks, and cattle) for thousands of years, since well before the time of Genghis Khan. Of particular interest to us—though we love all of God's creatures equally—are the goats, because they are cashmere goats. They have some of the world's warmest, softest hair (which is under their coarser, outer hair but not exclusively on their belly, just for the record), which used to be considered a true luxury item. It is rumored to have been a particular favorite of Empress Joséphine, who was given shawl after shawl by her husband, Napoleon (Was he compensating for something? We may never know.) but has now become, in some places, a relatively affordable luxury for the masses.[3]

When Empress Joséphine started wearing cashmere, no one else in Europe was. She was also pretty sure no one else ever would. In a letter to her son, Eugène, she wrote of her shawls, "They seem most ugly to me. Their great advantage is their light weight. I do not believe they will catch on; no matter, I like them for they are extraordinary and warm."[4] Clearly, Joséphine didn't get it—cashmere quickly became a coveted and relatively rare luxury item in the West, and it remained that way until the late twentieth century.

In the 1990s, everything changed, which is how we got to where we are now, where every fast fashion brand has a cashmere collection at prices that would have made Napoleon feel like he'd been ripped off. What changed? A few things. In Mongolia, the collapse of the Soviet Union and the subsequent privatization of industry generally meant that people could own their own herds and not depend on the state to set the price of goats or cashmere or hand out their wages. So Mongolian herders bred and bought more goats, increasing the supply of cashmere in order to make a living. By that time, China had already begun to open itself up to the West, leading to an industrial explosion, particularly in the garment industry.[5] Cashmere was no exception; people looking for work began to flood into Inner Mongolia (which is an autonomous region of China and not part of Mongolia) and other areas of the high grasslands to raise livestock and work in the new factories. From the 1950s to the 1980s, the population of Inner Mongolia tripled, from 7 million to 21 million people. (They had already begun to move into this area in the 1950s and 1960s, after some encouragement from Chairman Mao, encroaching on the nomadic herders' lifestyle and homeland to dramatic effect. Some migrants arrived on the Alashan Plateau in 1956 and set up the city of Wuliji, where they dug wells and built a wooden furniture factory. Within ten years, all of the trees were gone and the factory had closed.)[6] During the 1990s, when the government was urging industrial development across western China, the Japanese economy went into a recession. Japan had been among the biggest buyers of Mongolian and Chinese cashmere, so all of a sudden there was a lot of cashmere available for cheap and a new market to sell it to: America.[7]

It feels, on the one hand, slightly crazy that these giant political and financial movements have made seismic changes in the world of nomadic goat herding on the edges of the Gobi Desert, but here we are. In Mongolia, there were 5 million goats in 1990,[8] about 19 percent of all of the livestock there.[9] Now, goats are 60 percent of the Mongolian livestock population.[10] A precise estimate of the goat population is difficult, but it was thought to be around 26 million in 2004.[11] China and Mongolia provide about 90 percent of the world's cashmere (the rest comes from Australia, the US, Iran, and Afghanistan).[12]

Livestock, no matter what kind, create waste and produce greenhouse gas emissions (mostly methane), accounting for somewhere between 14.5 percent and 18 percent of the global human-caused greenhouse gas contribution.[13] And there are a lot of livestock beyond the cashmere goats: more than 1 billion cattle[14] and, at least in 2011, about 1 billion sheep.[15] Enteric fermentation (farting and burping, but upscale) makes up the biggest proportion of greenhouse gas emissions resulting from wool production.[16] There isn't a lot of data on the greenhouse gas emissions from cashmere production, but we can reasonably assume that the same is true (or perhaps even more true, since cashmere goats typically graze, whereas some sheep might eat feed, which requires energy, fertilizer, etc.) about Mongolian goat flatus.

But cashmere goats' impact, like that of all livestock, stretches far beyond their unique digestive processes. These goats are different from other livestock because of where they are, and even though they are herded by people, the landscape they live in and the resources they rely on are more extreme and wilder. Because of the environment they come from and have evolved to live in, cashmere goats won't thrive in most places; basically, they can live only on natural grasslands at high altitudes, which exist in only a few places that are especially vulnerable to climate change.[17]

When confronted, as they often are, I imagine, with a field full of grass, goats tend to go whole hog. That doesn't mean that they act like pigs, in the literal sense of the word *pig*. That is a confusing use of that phrase, I admit, but I wanted to use it. Call it poetic license; call this book poetry; call me Ishmael. What I mean is that they eat the *whole* plant, and they eat rather voraciously, hence, *hog*. Whole hog. That was a hard few sentences, but ultimately, totally worth it and we all got through it together. When these goats eat grass or any other plant, they eat the stalk and the roots, too, pulling up the whole plant and leaving nothing in its place. (Some other grazing animals just nibble at the stalks or leaves.) These goats also have really sharp hooves, which, in basically every single article I read about cashmere, are described as "stilettoes" or "stiletto-like," which seems a little haute couture for a bunch of goats, but this is the "Fashion" section after all. Either way, the uprooting of plants leaves the soil less stable and less able to retain moisture, and the goats' sharp fashion shoe hooves break up the soil, also destabilizing it. When the winds come, the soil, now left unrooted and not tightly packed, blows away, covering the grassland in desert, sending dust swirling from the desert to parts east.[18]

Climate change is also a factor here, bringing more frequent and severe droughts to the region,[19] which make it harder for plants to grow, and bringing more frequent and more intense storms to the region, too. Over the last seventy years, the average temperature in Mongolia has risen by nearly 4°F, compared to a global average increase of 1°F in the last century.[20] According to the Green Gold Project, an organization that focuses on preventing overgrazing, 65 percent of Mongolia's grasslands are degraded as a result of overgrazing and desertification, though 90 percent could be restored to grassland in the next ten years if the management practices change *now*.[21] According to the United Nations Environment Programme, there has been a 30 percent decline in surface water in Mongolia in the last fifteen years; 90 percent of that country is at risk of being desert (including the 17 percent that is already desert).[22] According to a World Bank study, about six hundred rivers and seven hundred lakes have disappeared. There has been a 34 percent decline in the number of plant species in the Gobi Desert and a 30 percent loss of forested areas compared to 1917.[23] It can be hard to determine what proportion of these changes are because of climate change (so they'd be happening anyway) or to what extent human-goat activity (beyond human activity's role in climate change) is responsible. A study found that human activity was a stronger contributor to desertification in this area of China than temperature increase from climate change, but both are involved, and as the planet warms, it will contribute more significantly to desertification.[24] In a few words: it will get worse.

But right now, the goats are directly implicated. Public enemy number one. And the problem with overgrazing goats—more goats than the ecosystem can support, clearly—is that, in addition to everything else I just said, they eat all the grass, everywhere, which then doesn't grow back, so then the goats don't have enough to eat. Undernourished goats produce hair that is coarser, which makes for worse cashmere that herders can't sell for as much. They also are keeping their goats alive longer, and older goats, particularly older male goats, have worse quality wool, too.[25] (There are other reasons for the overabundance of goats: the Mongolians no longer have to slaughter their goats to feed the Soviet army; China doesn't import enough Mongolian meat, and China is Mongolia's primary trading partner.)[26] But if the goats can't survive or the quality of their wool is worse, the herders might not be able to sustain their nomadic lifestyle—lots of articles tell stories of herders giving up and moving to the cities to work in factories—which is sad for those particular individuals but also for global cultural history because people have been living that way there for many thousands of years.

To put it more simply: more goats to meet increasing demand at some moments and a glut of supply of cashmere fiber at others drives down the price, so the herders buy or breed more goats, which means there is more grazing and therefore more desertification, so the goats are undernourished, which makes their hair coarser, causing the supply of high-quality cashmere to shrink, causing the herders to breed more goats to try to meet the demand for better cashmere, and on and on forever until, once again, the world collapses in on itself like a dying star.

If you don't care about the rapid demise of herder culture or the Mongolian steppe environment (does "In Xanadu did Kubla Khan / A stately pleasure dome decree" mean nothing anymore?), well, this whole situation is also worsening the quality of all that cashmere that we have loved since Empress Joséphine told us it was ugly.

It could mean your cashmere sweater has some yak hair in it, which can be soft but isn't pure cashmere, which is, in theory, what you want.[27] However, like all cheap fashion (relatively cheap in this case and cheap because, as we know, the environmental costs aren't paid), this cashmere is another signal of wastefulness—it doesn't last as long and pills more easily—but the low upfront cost makes it seem worth it, or affordable, even if, in a global, moral, environmental, and ecological sense, it's not. And to be sure, there is still a lot of really expensive cashmere out there, which might be of better quality (meaning longer fibers) but still more plentiful than in Empress Joséphine's day.

The desertification issues aren't limited to the steppes of Mongolia and China. When the dust storms from the deserts in the northern regions kick up, they send the dust across the former plains, now desert, toward Beijing and other cities on the coast. There, it combines with soot from the country's many coal-fired power plants, and in a toxic cyclone, they travel across the ocean to the western United States in about five days, particularly in the spring. In 2001, Asian dust accounted for 40 percent of the worst dust days in the western part of the US. Los Angeles experiences at least one extra day each year of smog, with ozone levels exceeding federal limits, because of Chinese factories making goods that they send to us.[28]

Nonetheless, a lot of scientific research and common sense suggests that outsourcing manufacturing has resulted in an overall benefit to public health in the US, and Chinese pollution is a much smaller source of pollution here than things like traffic or other domestic industries. It's easy to put a healthy portion of blame on China for a lot of the world's pollution and for continued greenhouse gas emissions (larger in number than the US, though our per capita emissions are higher), but we're really the ones responsible. Greenhouse gas emissions and pollution resulting from Chinese production is a result of frantic industrial activity to satisfy the fleeting whims (cheap cashmere, for instance) of those in the developed world, particularly in the US. We are, effectively, laying our emissions and pollution at China's doorstep and calling it their problem, or wringing our hands about the poor air quality in Beijing and the struggling herders on the creeping edge of the Gobi Desert: *How tragic, but that has nothing to do with me.* (Chinese consumers, increasingly, are buying what Chinese factories produce, but the majority is still exported.) And I can't sit here and say that the 400 million Chinese people who have risen out of poverty should have stayed there, because an improvement in the lives of hundreds of millions of people is a good thing, but it has come to the world expensively, largely because of us.

I don't mean to take the blame that I just took from China and put it on you. I don't blame you. I don't blame me. It is not the fault of the consumer that cashmere is cheap, and it's not wrong to want nice things or to buy them, sometimes. The opening up of all these industries—like making cashmere a less-than-luxury item—puts the responsibility for making the right choice on the consumer, and that's not fair. It's not within your control how some company sources and produces its cashmere, or the size of the herd that they got it from. That should be the corporation's burden—whether they pay more to source better or they pay for the associated down-the-line impacts—or governments should make sure they act responsibly. And that may make cashmere cost more (upfront, though the long-term health and environmental costs would be less). I don't want to say that not everyone should be able to buy cashmere, but I also think that we can't all have unlimited amounts of cashmere, if we want to live in a world that isn't spinning into a desert, in order to keep us swathed in cashmere at cheap prices because that's what we've decided is important.

All of these problems are connected, and the lives we live in one place, which are an accident of fate to begin with, are not separate and distinct from the lives in another. These are problems we all share. The dust from desertification from cashmere goat grazing and climate change and the coal pollution from factories making clothing from that cashmere blow over to the US in a karmic windstorm. It makes people sick in Los Angeles and Beijing; it makes a nomad settle in the city; it makes the world warmer. It's all part of the same problem, and it's not just cashmere. It's everything we wear and how we use it.

Fuel

Now that we are closing in on the end of our journey together, it seemed like the right time to talk about fuel by talking about the things powered by fuel. Basically, this is all going to be about electrical energy and transportation, since emissions from the energy and transportation sectors make up more than half of our national total.[1]

There's a reason I didn't start with fuel, though it is arguably the most important problem to solve. It's partly that it's complicated, but it's also because our conversations about climate change and the environment usually revolve around fuel. The way that we talk and write about it is pretty technical and abstract and, therefore, often kind of boring. Which is to say I didn't want to write about it that way, but I also didn't want you pick up the book, see stuff about fuel first, assume the book would be technical and boring, and decide not to read it. I am nothing if not a savvy businesswoman, evidenced by my decision to become a journalist writing about the world's most popular, easy-reading topic.

Seeing that the above is true, or at least, it is my truth, I wanted to figure out how to write about fuel in a different way. Generally, there is some awareness that burning fossil fuels emits carbon dioxide and other greenhouse gases, causing the earth's atmosphere to warm. But within all of that, there is a lot more going on. There are problems with fossil fuels beyond their global warming potential, which would be enough (and will certainly make up some of this section). Yet, the burning of fossil fuels has other significant impacts, some of which can be more immediate than climate change, and almost all of which we aren't paying enough attention to.

By looking at various topics under the heading of "fuel," all of the chapters will build to the same conclusion: the fuels we are using and/or the way we're using them (overconsumption, basically) are polluting our air, water, and higher levels of our atmosphere. But more importantly, they are, throughout that process, exacerbating inequality or further entrenching the unequal and unsustainable systems that have gotten us to this point.

We use a tremendous amount of coal to produce about one-third of our electricity. When we talk about coal, we usually think about global warming or coal mining. Largely, we don't think about the actual burning of coal and what that produces: ash, forming one of the largest industrial solid waste streams in the country. It can be recycled—made into concrete or wallboard, for example—but if it isn't, it will never go away. And it is made up of all the bad things that are in coal that are not burned out—heavy metals, other toxins—and those things have to go somewhere. For the most part, coal ash goes into manmade ponds or dammed-up portions of lakes and rivers, where it can and almost always does leak into the groundwater or rivers and lakes or, in more catastrophic situations, spill out in disastrous amounts. It's a particularly insidious form of pollution, because we don't see it: it doesn't puff out of smokestacks. And most Americans *really* don't see it: coal ash pollution disproportionately affects Black Americans, communities of color, low-income communities, and people living in rural areas. For many of us, it's easy to ignore, so we do.

Or take wood. In Europe, and possibly soon in the US as of writing, wood counts as carbon-neutral and renewable energy, because of an accounting loophole written into the Kyoto Protocol, one of the earliest international climate accords, which has persisted into the EU's renewable energy directive (the agreement that requires European countries to get a certain amount of their energy from renewables). Not only does wood release carbon dioxide and other greenhouse gases into the atmosphere when it's burned, but it is also a pretty inefficient source of electricity generation, so more of it has to be burned to get the same amount of electricity as coal. Plus, trees help take carbon dioxide out of the atmosphere, and cutting them down seems... counterintuitive. And it is! Yet the UK and European countries are using wood from trees and leftover wood waste to help them meet their renewable energy targets, not counting the carbon dioxide emissions from doing so, and some of their wood is coming from American forests, mainly in the Southeast. Deforestation anywhere is a problem, but in the Southeast, which is especially vulnerable to hurricanes, water pollution (from things like coal ash), and drought, removing these forests is making things worse.

It's not a surprise that air-conditioning uses a lot of energy, and as the world warms, we will need (or want) to use more of it. But it also uses chemicals that have a powerful heat-trapping effect and must be handled properly to avoid escaping into the environment and causing more warming. The international community has agreed to limit the use of these harmful gases and replace them with others, which are more expensive and less widely available. But the hottest places in the world have the least air-conditioning, and they will have a harder time getting it under this agreement. I'm not saying that they should ignore the agreement, but it is also hard to make the argument that developed countries should get to decide that people in poorer countries, suffering from the daily impacts of climate change, can't have the same air-conditioning that we use with reckless abandon. How do we in the developed world balance economic development and a higher standard of living for everyone, which together have meant historically more emissions, with the urgent need to fight climate change by reducing emissions, without engaging in some kind of climate paternalism?

And then there's transportation. As we move more and more of our stuff around the world—different parts of different things come from different places and are assembled somewhere else and then sold in stores on the other side of the world—how much fuel are we using, and how are we using it? When we transport things by plane, we are sacrificing energy efficiency for speed, in a big way. Planes are amazing, but they also use tremendous amounts of fuel. More stuff is being sent around the world on planes, and more people want to travel. How do we make that happen without destroying the planet? Can we? If we use cargo ships, trade becomes more efficient in terms of fuel use, but that only begins to capture some of the environmental impacts of this industry: the fuels used in industrial shipping can be particularly dirty and can accelerate the melting of Arctic sea ice. Scrapping ships and building ports destroys vulnerable aquatic ecosystems, which also are capable of sequestering a lot of carbon dioxide. People who live near ports are exposed to much higher levels of airborne pollution. And this comes back to the question of whether you can decouple economic growth (or GDP) from greenhouse gas emissions. (So far, science says: no.[2])

Cars and trucks pose similar problems, except that we use them a lot more, and we seem to want to keep using them as much as possible and then more on top of that. Automobile pollution is something we're all pretty familiar with, but in the US and other areas where this issue has been studied, communities of color are disproportionately exposed to and affected by transportation pollution. Researchers at the University of Washington found that, in 2010, nonwhite minorities were exposed to 37 percent more pollution in the form of concentrated nitrogen dioxide (a particularly harmful gas from the burning of fuel) than white people.[3] While some people are aware of environmental injustices—like toxic waste sites or industrial facilities being disproportionately placed near disadvantaged populations—even fewer are aware of the disproportionate impacts of everyday pollution, especially when we think we all deal with cars and their pollution. I also think that, even though we hear a fair amount about cars, we don't hear about this, so it was a way of looking at cars that felt different and new and important.

What will the future of transportation look like? Ride-hailing app companies are promising us a lot: fewer cars on the road, and the ones that are left are electric and/or drive themselves and we all just share them, because if there's something Americans love, it's sharing their stuff with strangers. How are these companies doing so far on these claims? And how does an electrified, automated future happen? Do we have to make it happen? Who gets to live in that future, and how does it affect the use of and investment in public transportation, predominantly used by lower-income communities? Are these apps the problem, or are cars the problem? (Yes, both.)

I hope I didn't give it all away. In fact, I know I didn't, because these issues, like the others we've looked at, are complicated and contingent on lots of variables. In this section, you'll also see some more solutions than you may be used to getting from me. In this area, more than any other, the changes we can make—perhaps as individuals, but more importantly as a collective—will have a really important and big impact, if we decide they are worth making. Obviously, I think they are.

The way we use fuel most powerfully demonstrates how closely we are all connected: how driving my car affects the health of people who live near the road, what impact the movie you stream on your computer might have on someone living in Ohio, how a ship crossing the Arctic Ocean could change the life of someone in Greenland, why the rose you bought at Walmart brought jobs to rural Colombia but results in the release of tons of carbon dioxide more than it otherwise might, and other connections I hope you will make when you read what comes next. More than anything, the conversation around fuel tells me that this is about all of us.

The Other Problem with Coal

When I first started reporting on coal ash in early 2017, I assumed I would not be interested in it, because I had never heard of coal ash. There's a chance, unless you are from the Carolinas, Tennessee, Montana, or a few other states or have deep and abiding ties to the coal industry or the cement industry, that you haven't heard of it, either. Now that I do know about it, I can't believe we don't talk about it all the time. To me (and some environmental lawyers and advocates I interviewed), coal ash is one of our most pressing environmental issues, and we are biding our time until the next disaster.

Coal ash, which is the by-product of burning coal to generate electricity, is one of the largest industrial solid-waste streams in the United States; we generate more than 100 million tons of it every year, producing some in almost every single state.[1] If that weren't enough, coal ash contains all of the bad things: mercury, lead, arsenic, boron, cadmium, chromium, lithium, selenium, cobalt, and more![2] (Many of these substances are known carcinogens, even in small quantities.) Since it's what's left over from burnt coal, it doesn't go away, ever (it doesn't biodegrade or decay), and here's what power plants do with it, mostly: they store it in water, dammed up along the banks of rivers and on the shores of lakes all over the country. If they don't store it in water, they store it in landfills, often uncovered, so "fugitive dust" (it's on the run!) can escape into the air, where it can get into people's lungs.[3]

We didn't used to produce so much coal ash. Before the Clean

Air Act and improved technology, a lot of what is now coal ash just puffed out of the smokestacks of power plants. That certainly wasn't good for the air, but now we have a mounting surface and ground-water pollution crisis—and some of the contaminants still get into the air anyway. About 40 percent of coal ash is recycled—made into concrete or wallboard, used as structural fill for abandoned mines, as a top layer on unpaved roads, spread on icy or snowy roads, or added to agricultural land. Some of those uses, when the ash is not contained, as it is in cement or drywall, can be problematic, and you'll soon see why. (Some environmentalists argue that the real recycling rate for "beneficent use" is only 20 to 25 percent.)[4] But the rest of it sits in landfills or in ponds just about everywhere.[5] And they truly are everywhere: there are at least 1,100 coal ash ponds spread across almost all of the states, plus another four hundred or so landfills.[6] According to the Southern Environmental Law Center, every major river in the Southeast is connected to at least one coal ash pond.[7]

Just in the last decade, there have been two major environmental disasters involving coal ash. In 2008, a dam at the Kingston Fossil Plant, a Tennessee Valley Authority power plant in Tennessee, broke, releasing over a billion gallons of coal ash slurry into the Emory River. The coal ash buried three hundred acres in toxic sludge and coated the river bottom for miles downstream with heavy metal pollutants.[8] (I'll come back to this one.) The second was in 2014, when an old pipe in a coal ash pond at a retired power plant in North Carolina ruptured, and 39,000 tons of coal ash and up to an estimated 35 million gallons of wastewater leaked into the Dan River.[9] Though much smaller in scale, the broken pipe let the coal ash water flow for six days before Duke Energy,[10] the utility that owned the plant, plugged the leak, by which time it had traveled seventy miles downstream.[11] In 2015, Duke Energy's subsidiaries pleaded guilty to nine Clean Water Act violations committed across the state.[12]

During Hurricane Florence, which clobbered North Carolina in September 2018, a landfill collapsed and a dam was breached at one Duke Energy power station, three coal ash ponds were flooded at another Duke site, and dams were also breached at a third site, flooding a lake with coal ash.[13] As climate change makes hurricanes stronger and more frequent, floods like those caused by Florence (and like Hurricane Matthew in 2016, which created similar coal ash problems)[14] become increasingly likely and increasingly threaten to release more coal ash into the environment: the water we drink, the rivers and lakes we fish and swim in, and the air we breathe.

But the disasters, major spills, and criminal violations of federal environmental law are the rare big-ticket items in the eternal auction of industrial pollution. Every day, coal ash is contaminating groundwater, lakes, and rivers all over the country, particularly in the American Southeast. For the most part, the responsible parties get away with it, because coal ash disposal of this kind is, by and large, legal.

I started learning about coal ash when I was a reporter at the *New York Times* and hunting around for a new story to work on. Another reporter told me about a million times that I should write about coal ash, that he really thought it was a major oversight that we weren't covering it. So I started calling some environmental lawyers and asking some questions—primarily "What is coal ash?"—and if there was anything going on that might be interesting to readers of the *Times*, which is how I found out about the Gallatin Fossil Plant in Gallatin, Tennessee. At the time, there was a trial going on in federal court in Tennessee—the first time a Clean Water Act case had gone to trial in the state—which alleged that the Gallatin power plant, also owned by TVA, was polluting the groundwater around the site. So I went down to Nashville, which is about thirty miles downriver from Gallatin, to meet with lawyers from Southern Environmental Law Center (SELC) and representatives from TVA and visit the power plant. Beth Alexander, Anne Davis, and Amanda Garcia, lawyers from the SELC, gave me a crash course in what had happened at that particular site, which, after four hours of them repeating themselves over and over again, I now know like the back of my hand. It's a little technical, but, broadly speaking, here's how these lawyers explained their case: the coal ash ponds sitting on the banks of the Cumberland River were releasing toxic chemicals into the groundwater through the soil, and even though the plant had a permit to release some contaminants from the pond into the river, some of the metals and the amount of them being released were illegal. Additionally, they demonstrated that from 1970 to 1978, one of the coal ash ponds essentially drained into the river, with everything put into the ponds escaping into the groundwater and the river, amounting to 27 billion gallons of coal ash slurry, more than one hundred times greater than the volume of oil that spilled into the Gulf of Mexico from the Deepwater Horizon disaster in 2010.[15]

But the daily, mundane problem—coal ash contaminants leaking into the groundwater—is the real problem with coal ash. Nearly half of the coal ash ponds in the country don't have a liner, sort of like a garbage bag, which protects the soil and groundwater from the pollutants; more than half don't have a way to catch what might leak out of them. Many of them are also built below the water table and are basically sitting in a pool of the water we might drink.[16] About a third of coal ash ponds are within five miles of a public drinking water intake; about 80 percent are within five miles of a drinking water well.[17] A study of groundwater next to coal ash ponds in five southeastern states found that every site had concentrations of coal ash contaminants higher than what is found in nature, providing strong evidence of leaking, whether the ponds were actively receiving waste or not.[18] The EPA has identified more than 250 instances in which steam electric power plants, often fueled by coal, have harmed or potentially harmed surface or groundwater.[19] Physicians for Social Responsibility, a doctor-led organization that mobilizes on behalf of public health issues, says that if you live near an unlined coal ash pond and you get your water from a well, you have a one in fifty chance of getting cancer, which is 2,000 times greater than the EPA's goal of reducing the risk of cancer from carcinogen exposure to one in 100,000.[20] So even if these ponds aren't built on riverbanks and lakeshores, threatening collapse and havoc and the dogs of war, they are still posing a major problem.

The next day, I went to visit the power plant. I don't know what I was expecting, exactly—as someone who grew up in Manhattan, I didn't have a lot of exposure to coal-fired power plants, which is part of the problem. The power plant itself met my expectations, but I only saw it from the outside. But there were mountains of coal, which I was told would last for a month, though their size (and my ignorance) made me think you could generate enough electricity to last until the sun burned out. The rest of the facility is what surprised me: what used to be a coal ash pond is a forest with deer running across it. Wind blew ripples across the surface of the ponds, sitting at the bottom of gently sloping hillsides, with a small stream babbling away, quietly swirling into the river before being rushed away. It was all very eerie: a controlled vision of nature, looking green and wet and alive on the surface, but covering up a particularly sinister type of waste. And it was hard for me to navigate the situation: having reviewed the trial transcript and analyzed the evidence—documents from both sides—I felt pretty sure that what was happening at Gallatin wasn't right. (I haven't even mentioned that the nearby City of Gallatin gets its drinking water from a site about a mile downstream from the power plant, and officials there see the coal ash seepage as a threat to their drinking water supply, though they had not identified any health effects. Plus, Nashville and other nearby cities also get their water from the river, so upwards of 1 million people rely on the Cumberland River for their water.) But the TVA officials (they had brought the president of the agency and chief engineer to meet with me, the cub reporter, which felt like a little much) were telling me that there wasn't anything to see here. It was hard to square all that I had learned with what I knew about TVA: a powerful federal agency with a presidentially appointed board, which had transformed much of the Southeast during the Great Depression, electrifying the region, bringing it closer to the future that the rest of the country was enjoying at that time, and providing jobs for a community that had been (and often continues to be) neglected.[21]

All of those dynamics are part of the problem. Studies by the EPA have shown that coal ash ponds are routinely sited near rural, low-income communities of color, which is part of the reason why the issue doesn't receive enough attention. In many circumstances, coal ash pollution affects powerless and poor communities; if it were in rich, white communities, Beth Alexander said, it would get a lot more attention.

By contrast, utilities in many of these states are practically all-powerful. I talked to Thomas Cmar, a lawyer for Earthjustice who works on coal ash issues, about this problem. "Some of the most powerful entities in our country are the ones that sell us power, and I don't think most people have a good sense of that," he said. When we talk about coal, we talk about why it's a problem in terms of "the coal industry," by which we usually mean mining in deep caverns far below the surface of the earth the rest of us stand on, blasting off the tops of mountains, and greedy coal executives and black lung disease and greenhouse gas emissions.[22] But so much of the power of the coal industry actually comes from these utilities, which are responsible for actually burning the coal that, even though its use is declining rapidly, is still our second-greatest source of electricity (after natural gas), producing about 30 percent of our power in 2017.[23] In eighteen states, coal is still the biggest generator of electricity.[24]

After I wrote about the case, a federal judge issued a pretty strong ruling in favor of the environmental group and against TVA, ordering them to excavate the ash ponds and move the ash somewhere else.[25] TVA appealed the case and also said that excavating the ash was too expensive, which, given that TVA is a federally-owned agency not beholden to shareholders and has been polluting the water for decades, seemed like a weak argument. But their appeal before the Sixth Circuit Court of Appeals in September 2018 was successful, reversing the earlier ruling.[26] A separate lawsuit making its way through state courts is, as of this writing, unresolved. While excavating the ash and draining the ponds is the best solution, it still poses problems because it has to go somewhere and most people don't want a giant coal ash landfill near their communities, even though storing coal ash in lined landfills with a cover is the safest and best option. After the Kingston spill in 2008, TVA carted the ash/sludge away to a dry and lined, though uncovered, landfill in Alabama, right outside a predominantly African American community, which fought against receiving the coal ash and filed a civil rights complaint against the landfill company for environmental racism and health consequences.[27] Then the landfill company sued some of the town's residents for defamation, if you can believe such a thing. Nine years later, the defamation case was settled and the town won some environmental protections, but the residents are still living with the health effects.[28] In March 2018, the EPA dismissed the residents' claims that their civil rights had been violated.[29] And then, the workers who came in to clean up the site of the Tennessee spill were not given any protective clothing or equipment and ate their lunches on top of mounds of coal ash despite the fact that it contains numerous harmful substances. A lawsuit against the engineering firm contracted to clean up the site claimed that thirty workers died of illnesses resulting from working at the site, and two hundred more workers got sick there.[30] A federal jury ruled in favor of the workers and their families in November 2018, who can now seek damages from the engineering firm.[31]

How is it possible that this is still a problem? It's a good question. Until 2015, coal ash itself had never been specifically regulated by the federal government, though under the Clean Water Act, everyone, including power plants, had to get permits from state governments if they wanted to let a little toxic waste get into a body of water. The permits, in theory, set limits on how much and what kind of contaminants could escape. At no point, however, were the contaminants supposed to be able to leak out into the groundwater.

It sounds like a logical, airtight, and watertight system in which nothing could go wrong, right? Let's allow powerful monopoly utilities to wash their industrial waste into our lakes and rivers— shouldn't be a problem, since they are generally responsible and want what's best for us, regardless of profit...

My righteous indignation aside, it became a system ripe for abuse, even though the Clean Water Act is a good and strong environmental law. It became abused because of how powerful utilities are and how willing state governments can be to cooperate with them.

In 2015, the Obama administration put the first ever regulations into effect relating to coal ash waste. They weren't as strong as environmental groups wanted, but they did make a difference: leaking coal ash ponds had to be fixed or closed, utilities had to monitor their groundwater and give the information to the public, citizens could sue utilities that were violating the rule, among other important provisions. Environmental groups sued to make it stronger; lobbyists for utilities did the opposite.[32]

When Scott Pruitt became the EPA administrator in 2017, he decided he would rewrite those rules. When Andrew Wheeler, Pruitt's replacement and a former coal lobbyist, became acting administrator of the agency, the first rule he signed was the revision of the coal ash rule, phase 1 (phase 2 is coming).[33] Almost needless to say, it's much weaker and gives more enforcement power back to the states. "We saw what the world looked like when states were responsible for permitting these facilities," Thomas Cmar said. "The whole point of the rule was that there needed to be a set of minimum standards because states did not have the technical ability or the political will to adequately monitor groundwater around these sites."[34] That's the world we're living in again.

So why should this matter to you, or what does this have to do with your environmental impact, allegedly the subject of this book? First, I believe that it is possible—nay, important—to care about things that don't necessarily affect us personally or things that we don't feel we are personally responsible for. Part of the goal of this book is to explore how we are all in this together, how the systems we participate in affect all of us, even if not directly. And we all do participate in this particular system, even if we don't live in a place where coal is still king. If you flash back to the part about data centers, you may recall that 70 percent of global Internet traffic passes through Loudoun County, Virginia, or that many of Amazon's data centers are in Virginia and Ohio, where 30 percent of the electricity is from coal. If the renewable facilities that Amazon has built or invested in don't meet the demand, the electricity could be coming from coal. So when we watch Netflix in New York, we might be, in a way, creating a demand for coal to be burned somewhere else, and that makes some coal ash, which could end up in someone else's water. I'm not saying that watching

Netflix makes you an industrial polluter or a bad person, because I don't think that. I think that it's just so easy for us to be disconnected from the consequences of our consumption. To me, thinking about coal ash shows how closely we are all connected and that we all have a vested interest in solving these problems. And that starts with understanding how this works, because for those of us who had the luxury of not knowing what coal ash is, our lack of awareness is part of the problem—that's how these problems start and how they become so entrenched as to seem unsolvable.

Now we'll look at another energy problem that also has significant consequences for the American South.

The Wood for the Trees

There are so many things to give the British credit for: scones with cream and jam, Shakespeare, the idea that no one is above the law, ruthless colonialism, warm beer, I could go on. But the thing they did that transformed the world more than all others is to burn coal for energy. It is almost impossible to imagine our world without coal; perhaps there would have been a different kind of Industrial Revolution and we would have ended up in the same place, but because of the steam engine and the relative efficiency of powering it by coal, here we are.

But I'm not talking about coal right now. Or, I am talking about it because I'm talking about Great Britain. In 2017, the British government announced that it would require coal-powered electricity generators to use expensive technology that limits carbon emissions or else close by 2025;[1] they, at that time anyway, were also subject to the European Union's Renewable Energy Directive, which required that member states get 20 percent of their energy from renewable sources by 2020 (and cut greenhouse gas emissions by 20 percent and increase efficiency by 20 percent, too).[2]

In line with that proposal, Drax, the operator of Britain's largest coal-fired power plants, announced that it could stop using coal by 2020 and switch over to renewable energy and gas-powered plants.[3] There are also periodic fanfares when England has gone twenty-four hours without using any coal at all.[4] Huzzah! All hail!

Prince George is the future king! We get it, it all sounds good. But how good is it?

It's not good. I mean, it could be good, but it's not.

Why not? Drax, along with many other coal-fired power plant operators across Europe, has already converted some of its boilers (and plans to convert the rest) to be able to burn wood pellets (also known as woody biomass and often made from wood waste) instead of coal.[5] In Europe (and soon in the US),[6] this counts as carbon-neutral (or zero-carbon) renewable energy.[7] The same as wind and the sun. I guess it's renewable in that you can plant new trees, except not, according to almost all scientists, environmental lawyers, and conservationists, some of whom call this convenient (for power companies) feature a "critical climate accounting error."[8] Burn!

How did it happen? It all goes back to the first Intergovernmental Panel on Climate Change (IPCC) science assessment in 1990, which counted emissions associated with forestry as emissions from changing the way land is used (land-use emissions is what they're usually called), even if the harvested wood was used for energy. (They didn't count the energy emissions because they were worried about counting emissions twice. As if that has been our problem. Just kidding—scientific accuracy is important, I know.) And voilà, the loophole. It was further codified when the EU established a carbon cap-and-trade program in 2005 and listed emissions from biomass, which covers most wood pellets and wood waste as exempt from cap-and-trade fees. Now the European member states use the IPCC accounting system and don't count emissions from burning the wood and don't count the emissions from harvesting the wood for energy. Another assumption in this creative emissions accounting solution is that when new trees are planted to replace the ones that were harvested and meet the demand, they will absorb the carbon dioxide at the same rate that the old ones did. And in the EU, all of that makes wood "carbon neutral," though countries do have to count the emissions associated with producing and transporting the pellets.[9]

In both the IPCC assessment and the EU'S renewable energy directive, large-scale electricity generation from wood did not seem like it would be a big thing. For a long time, it wasn't. But that was before. Now, as Europe prepares to meet the 2020 goals, they are ramping up their renewable energy portfolios, and now get just under half of their renewable energy from wood, though energy from biomass can come from other sources.[10] Currently, the UK is the world's largest importer of wood pellets.[11]

There are a lot of problems with this. *Climate Central* reporter John Upton notes that an "analysis of Drax data reveals that its boilers release 15 to 20 percent more carbon dioxide when they burn wood than when they burned coal," largely because wood (even in pellet form) is less energy-dense than coal, so more of it is needed to produce the same amount of energy.[12] Second, as we learned in the section on corn (Remember corn? Don't you miss it?), when you cut down plants of any kind, you release carbon that was sequestered in the soil. The carbon is still sequestered in the wood itself, but it gets released when the wood is burned. But now, perhaps you tell me, we can plant new trees to take up the carbon dioxide emissions, then boom, neutralized. Maybe, except that studies have shown that it would take decades, if not centuries, to compensate for the sequestered carbon lost with deforestation. In the meantime, there's a lot of excess carbon dioxide in the atmosphere that will cause a lot of warming before your new trees have fully paid back that carbon debt.[13] But, because wood counts as a renewable energy in most countries, the power plants that burn it receive tons and tons of government subsidies and avoid the fees that come with emitting massive amounts of carbon dioxide.[14]

Another big problem is where the wood is coming from. Parts of Europe, particularly the Nordic countries, have a lot of wood.

But it doesn't have enough wood to meet its needs—about two-thirds can be sourced from the continent, but the rest is coming from other countries: a little from Canada, Russia, Ukraine, Belarus, and other countries, and almost half of its imports come from the United States.[15]

Between 2011 and 2015, exports of American wood pellets more than doubled, and most of the wood came from forest and pine plantations in the American Southeast. Across the Southeast, twenty-seven wood pellet mills had been built by 2015, and at least twenty-five more are being planned in order to meet Europe's insatiable demand.[16] Traditionally, wood pellets were used for heating and made up of either wood left over from forestry operations, or by-products from other wood industries, like sawdust. Pellet companies claim that that's what they're doing, but investigations by environmental groups and others have shown that they're not. Instead, investigations have found that whole trees are being used to create the wood pellets, and not just trees from pine plantations, ostensibly grown for industrial purposes, but also hardwood from old-growth forests that are up to a century old, all across the South. These trees, which have absorbed and sequestered carbon for decades, are now being cut down and burned. Plus, they take longer than pines to recover the same amount of carbon dioxide, meaning that the carbon debt is paid back much more slowly. Enviva, the world's largest producer of wood pellets, now with mills in North Carolina and Virginia, uses hardwoods supplied by loggers to create 90 percent of its pellets, according to the *Climate Central* investigation.[17] An Enviva spokeswoman told the *Wall Street Journal* that it periodically audits its loggers and the company believes that the wood material that is sourced from hardwood forests and wetlands is logged sustainably.[18]

Why is this happening in the South and not, let's say, the timber-rich Pacific Northwest? Mills there shut down in the 1990s when owl habitats received federal protections.[19] Many states in the South lack environmental protections for forests, plus the vast majority of forests there are privately owned, so regulation and enforcement is more difficult.[20] But no matter why it's happening, it's happening at an alarming rate: a study showed that from 2000 to 2012, logging was four times more disruptive in the forests of the American South than in the rainforests of South America, directly coinciding with this period of rapid growth of the wood pellet industry.[21]

It's not just the carbon that escapes from the soil when these trees are cut down, or the loss of future carbon sequestration, or the wholesale destruction of native and biologically important ecosystems in favor of corporate interests, though that might already be enough for you. But it is not enough—it is never enough—for me. My love for disappointing and depressing facts, as you perhaps have come to learn, is endless; it is a love story of generations, one that moves mountains, ascends to the mountaintops, and then explains why those mountaintops now appear flat because of something called mountaintop-removal mining.

Pellet production has its own environmental impacts: it uses up energy (to dry the wood, chop it all up, dry it again, chop it up again, and then press it into pellets), therefore releasing carbon dioxide, and the process also causes thousands of tons of other harmful gases to escape into the atmosphere. In the drying stage, the burning wood emits fine particles, carbon monoxide, and nitrogen oxides, all of which have effects that are harmful to human health or the environment or both. The drying wood also releases volatile organic compounds, which chemically react under sunlight and produce ground-level ozone, a major component of smog. The government is largely not keeping track of these emissions, many of which are potentially illegal. According to a report from the Environmental Integrity Project (EIP) that looked at twenty-one wood pellet plants in the South that export to Europe, seven violated their permits by releasing pollution in excess of the legal limits, while another four had permits from state governments that violated the Clean Air Act by allowing these plants to forgo the required pollution control equipment. More than half had pollution above the legal limits or didn't have adequate pollution control technology. There are other considerations, too: of the fifteen largest facilities that the EIP looked at, eight have had fires or explosions since 2014 in five different states, releasing large quantities of air pollution. A fire at a pellet storage site in Port Arthur, Texas, burned for two months; one person was killed during the cleanup, and dozens of local residents sought medical treatment for exposure to smoke. If Europe's demand for wood pellets continues to grow, it is increasingly important that the dozens of new pellet plants (as well as the existing ones) operate within the legal limits.[22]

And it's not just Europe anymore. In late April 2018, Scott Pruitt announced, alongside representatives from the forestry industry, that the US would also treat energy generated from wood pellets as carbon-neutral. In the US, currently only about 2 percent of electricity was generated from wood and wood waste in 2017, most of which come from residue from logging, manufacturing, and discarded timber, but that could change.[23] If emissions from biomass were counted, this energy source would still pose problems, but at least we'd know what we were dealing with.

I would be wrong, though, if I didn't mention that a sustainable forestry industry is important to keeping forests healthy. As we increasingly inhabit every crevice of the country, we naturally encroach upon forested land. Because we now live near forests, we don't want them to burn, so many states and the US Forest Service practice fire suppression. But many scientists and conservationists argue that forests need some fire to stay healthy: it burns up the uncleared dry and dead wood, allowing space for new trees to spring up or to reach higher into the forest canopy, and it helps to spread seeds.[24] If a forest has a lot of dry and dead wood, fires can spread more rapidly and become much more dangerous,[25] which helps explain why fires in the American West have become so damaging (climate change explains much of the rest of it).[26]

Forests that don't have fires are also crowded with trees, which can make it easier for pests to spread across a forest system and have more devastating consequences.[27] It makes sense that we don't want fires, so that means that we need to make sure that forests stay healthy and that dry and dead trees are removed. Sustainable forests store more carbon, too. A market like the European wood market, huge and growing, could create an incentive to keep forests unsustainably managed.

It's also important to pay attention to the fact that this is happening in the South. Anywhere in the US, it would be important, because forests absorbed nearly 12 percent of our carbon dioxide emissions in 2016.[28] But the American Southeast has incredible biodiversity, with more plant and animal species than many other regions in the country, but with minimal environmental protections.[29] As climate change disproportionately affects the American South,[30] forests are even more important: they can provide protection during hurricanes and soak up floodwater.[31] They protect drinking water supplies at a time when drought, overirrigation, and subsidence are cutting into our drinking water supply and making it dirtier. The American traditions of environmentalism and conservation have their origins primarily in the northern and western parts of the country; the South has often been ignored, despite the immense natural wonders and environmental challenges there. There has not been a big tradition of environmental history in the South, and in my lifetime, the South has been home to more environmental tragedies and exploitations than triumphs. All over the country, marginalized people have been exposed to more environmental hazards than their wealthier, whiter neighbors, and that phenomenon of environmental injustice can help explain what we're dealing with here. These problems are all connected—the problems in the South are a problem for people living in the Northeast or in the Pacific Northwest. The one good thing about climate change (and that is a highly, highly qualified use of the word *good*) is that it might finally make us understand that we are all in this together, that the problems of one group of people are everyone's problems, and that we all breathe the same air and inhabit the same small planet.

Except when we don't, which brings me to air-conditioning.

Staying Cool, Getting Hotter

"It is likely the single most important step we could take at this moment to limit the warming of our planet and limit the warming for generations to come," John Kerry, then secretary of state, proclaimed at international climate negotiations in October 2016.[1]

If you had to guess what he was talking about, would you have guessed air-conditioning? Well, that's exactly what he was talking about, it turns out.

More specifically, he was talking about an international agreement reached in Kigali, Rwanda, to limit and eventually phase out hydrofluorocarbons (HFCs), a chemical used in refrigeration and air-conditioning, which is an incredibly powerful greenhouse gas—able to trap thousands of times more heat, for the same volume, than carbon dioxide.[2]

He also wasn't exaggerating: according to the authors of *Drawdown: The Most Comprehensive Plan Ever Proposed to Reverse Global Warming*, a book that ranks one hundred of the most effective solutions to end global warming, refrigerant management comes first; other research suggests that improving efficiency and phasing out HFCs from air-conditioning and refrigeration could help us avoid as much as 1°C of warming by 2100.[3] Given that it's going to be pretty hard to avoid 2°C of warming, this seems like it could be a big deal.

But HFCs and refrigerant management are only part of the story. As with so much in this book, a lot of the problem comes down to how we use whatever thing we happen to be talking about at the moment. With air-conditioning, it's both the heat and the humidity. Not really, but also yes, really.

The problem around HFCs is easier to explain, so I'll start there. The refrigerant industry started using these chemicals in the late 1980s, after the signing of the Montreal Protocol in 1987. Under that agreement, the international community agreed to phase out and eventually ban substances that deplete the ozone layer, like chlorofluorocarbons (CFCs) and hydrochlorofluorocarbons (HCFCs), which were used in refrigeration and air-conditioning and had been shown to cause significant damage to the ozone layer.[4] At the time, HFCs were seen as a safe replacement because they left the ozone layer alone, and their global warming potential (how much global warming they can cause compared to carbon dioxide) wasn't known.

HFCs make up a relatively small portion of greenhouse gas emissions, and compared to carbon dioxide, they break up relatively quickly in the atmosphere, but they're incredibly powerful while they're there. And their atmospheric concentration is currently increasing by 15–20 percent every year.[5]

If your air-conditioning works properly, it's probably not releasing these chemicals into the atmosphere. Some HFCs are released during the production process; if your air-conditioner has a leak, or if you dispose of it improperly, some molecules can escape.[6] (If you visit energystar.gov, you can get information on how to get your unit to the correct recycling facility.)

That's why the international community wants to phase them out. During the negotiations, there were some complicating factors: while the US and other developed countries wanted to *freeze* (get it?) the use of HFCs and dramatically lower their use over the next twenty years or so, representatives from the developing world—India, in particular, but also other countries—wanted a more gradual phase out, since millions of people in India are newly or just about to be able to afford air-conditioners that use HFCs. (Their replacements are more expensive and not as widely available, though they do exist.) Also, it's getting hotter in India. In the end, developed nations, such as the US and countries in the European Union, agreed to freeze levels of production and consumption in 2018 and reduce them to about 15 percent of 2012 levels by 2036. Other countries—China, Brazil, African countries—will freeze their use by 2024, and reduce to 20 percent of 2021 levels by 2045. India, Pakistan, Iran, Saudi Arabia, and Kuwait have a little more time: capping HFC use by 2028 and winding down to 15 percent of 2025 amounts by 2047.[7]

The deal isn't as strong as environmentalists wanted, nor will it avoid the additional 1°C of warming, but it's something.

Why doesn't it end there, with a mitigated but ultimately relatively happy ending? It doesn't end there because of the way we have used and may continue to use air-conditioning. I told a friend I had learned some really interesting things about the history of air-conditioning. She said that sounded like a parody of a boring NPR program that everyone would turn off. We'll see about that now, won't we?

Air-conditioning, as you also may have noticed from your electricity bill, uses a lot of electricity. Of the 1.6 billion air-conditioning units in the world, more than half are in China and the US, and keeping them running uses two and a half times more electricity than all of Africa uses in a year.[8] Nearly 90 percent of American homes have at least one air-conditioner. There are a lot of conflicting statistics from government data about how much of our electricity is used on air-conditioning, but it seems to be around 6 percent of all electricity in the US, and about 18 percent of residential electricity use.[9] Air-conditioning our homes causes more than 100 million tons of carbon dioxide emissions every year, and air-conditioning–related emissions account for approximately 8 percent of the GHG emissions from cars and light trucks, producing around 85 million more tons of carbon dioxide emissions.[10]

And we've basically engineered a system where we're going to use as much air-conditioning as possible in the US. Beginning after World War II, when residential air-conditioning became more widely available, Americans started moving to the Sunbelt and the Deep South, both of which are hot.[11] By the mid-1970s, these were the fastest growing parts of the country, mostly because Americans were moving there from somewhere else and bringing air-conditioners with them. In 1950, 500,000 people lived on the five-hundred-mile coast of the Gulf; now it's home to more than 20 million people, nearly all of whom have air-conditioning.[12]

Air-conditioning has also had some interesting cultural impacts: some historians argue that this air-conditioning–powered migration has been a significant advantage for the Republican party: air-conditioning allowed older people to move to the South and the Sunbelt, and white Americans adopted air-conditioning first, creating a northern clime in the south for themselves (at least indoors), and those demographics tend to vote Republican.[13] And they did: in pretty much every presidential election since 1950, the Sunbelt and much of the South has gone red (not from the heat).[14] There are some confounding variables here, but it's interesting to think about how technology influences everything.

It doesn't totally make sense that a lot of elderly people, who are the most vulnerable to heat-related illness and death, would move to the hottest parts of the country, which are now only going to get hotter, meaning they will need to use their air-conditioners more. But that's another problem with air-conditioning: it has encouraged settlement in places where there perhaps shouldn't be lots of people, and then caused them to use more and more air-conditioning. But now they're there, and air-conditioning makes living there possible. Since 1960, heat-related mortality has declined by 75 percent in the US.[15] Less crucially, a recent study of students at Harvard showed that cognitive function declined during a heat wave among students who lived in non-air-conditioned dorms compared to their cool peers.[16]

In addition to changing where people live, it has changed how we live, too: because we had air-conditioning, people stopped building homes with how to keep them cool in mind, forgetting about natural coolants like higher ceilings and more windows. Now, homes are designed for central air-conditioning, not ventilation. And homes are bigger: from 1953 to 2006, the average home size tripled and, correspondingly, so did air-conditioning use. Between 1993 and 2005, total residential electricity consumption for air-conditioning nearly doubled. The same is true for office buildings, which are cheaply built glass boxes that are easy to build and cool, but whose windows don't open, locking us into an endless cycle of heating and cooling and, if you're a woman working in an office building, you keep a parka inside or buy a space heater for under your desk, so you're heating and cooling at the same time and feeling completely insane.[17]

And it's getting hotter. Taking the period from 1951 to 1980 in the Northern Hemisphere as a base average, it's gotten a lot hotter. In the Northern Hemisphere, from 1951 to 1980, one-third of summer temperatures were average, one-third were hotter than average, and one-third were cooler. Since then, it's become even hotter. From 2005 to 2015, two-thirds of summer temperatures were hotter than average; 15 percent were so much hotter than average that scientists had to add the completely new category of "extremely hot."[18] And nights are warming faster than days: overnight low temperatures have increased in the US by 1.4°F in the century (plus a few decades) since 1895, compared with a 0.8°F increase for daytime high temperatures. Higher nighttime temperatures can make heat waves more fatal, because the body doesn't really get a chance to cool down.[19]

All that means we'll be using our air-conditioners more. It means more demand on the electrical grid, which can cause more blackouts. In some parts of the US, air-conditioning can account for more than 70 percent of peak residential electricity demand on very hot days.[20] It also means more fossil fuel use: there currently isn't enough renewable capacity to power our air-conditioners, and when demand increases, the gap is most likely filled by fossil fuel power. So that, in turn, means more of the harmful greenhouse gases that cause climate change. Here's a ridiculous thing: because air-conditioners vent hot air out of your house, if lots of people have them on during a hot night in a high-density city, the air-conditioners themselves can raise the overnight temperature outside by more than $1°C$.[21]

Especially as the world warms, more people all over the world will want air-conditioning, and more people everywhere are getting closer to being able to own units. By 2050, the number of air-conditioners worldwide is expected to shoot up from 1.6 billion today to 5.6 billion.[22] If air-conditioners in 2050 work like the ones we have now, they could use as much electricity by then as China currently uses in a year, resulting in more carbon dioxide emissions, too.[23] Meeting the electricity demand of air-conditioners accounts for 21 percent of the growth in global electricity generation.[24]

In the hottest parts of the world—Africa, Southeast Asia, Latin America, and the Middle East—home to 2.8 billion people, only 8 percent of the population has air-conditioning.[25] As incomes in those places rise, more people buy air-conditioners (income is a better predictor of air-conditioning purchases than a warming climate), and much of the growth will come from India, China, and Indonesia, which are all very hot and humid and, at the moment, mostly powered by coal.[26] (As incomes rise, people also buy other household appliances, like refrigerators—which also use HFCs and lots of electricity—and televisions, which generate heat and make air-conditioners work harder for the same amount of cool.)[27] Most of the air-conditioners sold in India today use twice as much electricity to produce the same amount of cooling as more efficient units. (Not to mention that air-conditioners in the European Union and Japan, which have similar rates of ownership as the US, are generally 25 percent more efficient than the ones sold here and in China.)[28]

We are headed for a cooling crisis. But it would seem colonialist and patronizing if more developed countries like the US, EU member states, and Japan were to tell these developing countries that they shouldn't get to have air-conditioning because it would use too much electricity, as we sit here, cool and comfortable, basking in the artificial breeze of privilege. And we, in the developed world, have gotten used to air-conditioning, which makes it harder to live without. Meanwhile studies have shown that people in developing countries need air-conditioning too, because it's getting hotter there, and vulnerable people—the elderly, the sick, and the young—are suffering.

Unlike lots of the other problems I've told you about, this one has a solution. Get up, everybody, and sing! If stringent efficiency standards are adopted, the average energy efficiency of air-conditioners around the world could more than double between now and 2050, reducing the demand on electricity generation and potentially cutting carbon dioxide emissions substantially.[29] If paired with increased renewable energy on the grid, it could cut carbon dioxide emissions related to cooling by 13 percent compared to 2016.[30] Buildings could be designed and built more efficiently, too, with better ventilation and fewer heat-trapping materials. If electricity prices better reflected their full cost—if, say, carbon dioxide emissions were factored into the cost—there would be more incentive for devices like air-conditioners and buildings to be more efficient. And if the new, less-heat-trapping chemicals are used, the air-conditioners that do run will have a smaller global warming effect.

In the end, you don't have to be hot—it's what's on the inside (of your air-conditioner) that matters. Consider this my retirement from comedy.

The Great Big Cargo Route in the Sky

A rose is a rose is a rose, and by any other name would smell as sweet, etc., etc. The thing about a red rose, especially one packaged up, caressed with white flecks of baby's breath, and untouchable beneath a glint of plastic wrap, is that we all know exactly what it's supposed to look like and what it's supposed to mean, which is love. For something so central to the vernacular of love and romance, even in their cheesiest forms, we know very little about where all of these roses, peeping out of plastic tubs at every corner store, Walmart, or flower shop, come from. Unfortunately, it falls to me to break your heart just a little by telling you.

Let's imagine for a second that you forgot Valentine's Day, and when you suddenly remember, all you can think to buy as a gift is a bunch of red roses. If you buy one of those rose bundles for your beloved, most likely it has come to you on a plane from Colombia, probably by way of Miami. The majority of roses Americans buy and give each other on Valentine's Day—around 200 million every year—are from Colombia.[1]

Every day of the three weeks leading up to Valentine's Day, thirty cargo planes fly from Bogotá to Miami, each carrying 1.1 million roses. Walmart alone, the biggest flower retailer in the country, purchased 24 million Colombian roses for Valentine's Day in 2018. From 2009 to 2015, flower imports from January 1 to February 14 increased seven times, while production of American roses, overall, fell by 95 percent, from 545 million roses to fewer than 30 million. The *Washington Post*'s Damien Paletta writes that this is largely a result of the 1991 Andean Trade Preference Act, which lifted export duties on goods from Colombia, Ecuador, Bolivia, and Peru as a way to move those countries away from cocaine production. Colombian roses kept their preferential treatment in 2012, when Congress passed the US-Colombia Trade Promotion Agreement in exchange for tariff-free exports of American corn, wheat, soybeans, and oil. (It always comes back to corn and soybeans, doesn't it?)[2]

In 2017, 4 billion flowers (roses and others), weighing 200,000 tons total (about one-tenth of a pound per flower), were flown from Colombia to the United States, representing about 40 percent of the total airline-reported payload (the combined weight of passengers and cargo) between the two countries. Carrying all of those flowers (and just the flowers, not including their packaging) used 114 million liters of fuel and released 360,000 metric tons of carbon dioxide, according to an estimate from the International Council on Clean Transportation. When the roses land in Miami, they might be sent elsewhere: another city in the US, Canada, Russia, and Japan are some of the major destinations. If they go to Japan, it causes the roses' carbon intensity to skyrocket (mixed metaphor, I know, but skyBoeingDreamliner lacks the same vivid imagery) by about 600 percent.[3] In total, 6 billion Colombian flowers are exported every year to ninety countries. These figures don't include emissions from the refrigerated trucks that take the roses to and from airports and warehouses, or any subsequent flights (except Japan), or the emissions from keeping the warehouses where many of the roses are rewrapped or assembled into bouquets at a cool 40°F, cold enough to keep the flowers from blooming but warm enough so the workers don't freeze or quit.[4]

This is not to say that American roses would have no carbon footprint, and that growing roses on an industrial scale would pose no problems if it all happened in the US. Packaging the roses for relocation or re-export has replaced some of the jobs lost in the collapse of the American rose industry (though not necessarily for those who worked on the farms) and has helped provide jobs and security for 130,000 Colombians.[5]

This is more to point out how things have changed: how many more things are being moved around by plane, even if it's still not that much, relatively speaking; how much we're all flying; and what all these airplanes circling overhead are doing to the planet.

And it's not just roses. Let's look at what happens over an average twenty-four hours in the wild world of international air freight: 100,000 takeoffs; 20 million parcels delivered; 1.1 million smartphones transported; more than 200 race horses flown (seriously); 6,849 vaccines delivered. About 1 percent of global cargo by volume travels on planes, but it represents 35 percent of global trade by value.[6] Lots of global cargo travels on passenger planes, but some, like millions of our Colombian roses, travels on dedicated freight planes, and the global freight fleet is projected to double in the next twenty years.[7] As we demand more of our stuff more quickly (hello, e-commerce, my old friend), more air freight won't be perishables and racehorses—it could be the new Fitbit you needed, like, yesterday.

Most of the growth, though, comes from us moving around. There are currently 20,000 planes droning on and on overhead, which is expected to grow to 50,000 planes by 2050.[8] There were almost 4 billion passenger trips on airplanes in 2017; that number is also expected to double in the next twenty years, with much of the growth coming from the developing world.[9]

That was all a very long prelude to saying that even without the growth, airplanes are big obstacles in the fight (or flight. I'll see myself out . . .) against climate change.

At the moment, airplanes account for at least 2.5 percent of global greenhouse gas emissions, about the same as Germany.[10] It's not surprising that you need a lot of fuel to get giant machines full of people and their bags and mail and flowers and racehorses and green beans and vaccines up into the air and make them stay there for thousands of miles.

Since 1970, global air traffic has doubled roughly every fifteen years;[11] the air cargo fleet is expected to increase by 70 percent over the next twenty years. A silver lining is that airplanes have gotten a lot more efficient over time—about 45 percent between 1968 and 2014[12]—as a result of better engines and design. And airlines adding more seats and cutting back on everything else—luggage allowances, ice in your drinks, etc.—means they're carrying more stuff on the same size plane.[13]

If airplane emissions continue to grow at their current rate, they are projected to triple by 2050.[14] (I'm focusing on international aviation because it forms the majority of emissions and for another reason which you will soon know.) And carbon dioxide is just one of the things they emit; they also spit out nitrogen oxides, sulfur dioxide, water vapor, and aerosols. Nitrogen dioxide, while not a greenhouse gas, has powerful heat-trapping qualities and also can cause ground-level ozone, which is dangerous for human health. Sulfur dioxide can have a moderate cooling effect (as can water vapor) and lead to the formation of cirrus clouds and contrails, which also can have a warming effect. Add it all up, carry the one, and yeah, it's still not great.[15]

Climate change itself will make air travel more difficult. Already, airports in places like Arizona are seeing cancellations and endless delays of flights on the hottest days, when the air on the runway is too thin to allow some planes carrying their maximum payload to take off.[16] A study in 2017 predicted that by 2050, one-third of flights scheduled to take off in the hottest parts of the day won't be able to do so at weights that make flying even somewhat economically sensible.[17] Changing air patterns, another result of warming global temperatures, will make flights more turbulent. So your flight will be delayed and if you ever take off, you'll barf—a vision for the world we can all get behind.

Emissions from aviation have proven to be pretty difficult to rein in. It's not hard to see why: unlike in the electricity sector, there aren't really a lot of low-carbon substitutes for jet fuel and the entire fleet of currently existing planes would have to be replaced in order to use the alternative jet fuels that are available. That plus demand—more people want to fly more of the time, especially as the global middle class grows, particularly in Asia— makes it even harder.

Then there are the political reasons. According to the International Civil Aviation Organization (ICAO), an agency of the UN, international airplane emissions (unlike domestic ones) don't belong to any one country. They belong to all of us. How lucky we are to live in such a generous world! The downside: this means that airplane emissions were not included in the Paris agreement in 2015; instead, it was determined that the ICAO would come to its own agreement, which it did. About a year after Paris, 190 countries signed on to a voluntary agreement to cut back on airplane emissions. Their plan: beginning in 2021 and voluntary for six years after that, cap emissions at whatever levels they reached in 2020.[18] Except not really—for any emissions over those levels, which will still be substantial, airlines will purchase credits to offset them, in the form of renewable energy generation, forest conservation, and other projects that prevent additional emissions.[19] ICAO also assumes that airplanes and airlines will further reduce emissions by 2 percent every year through efficiency improvements[20] and save about 2.5 billion tons of carbon dioxide emissions over the first fifteen years of the program.[21] Some experts who think the plan doesn't do enough to help the world meet the 2°C limit set by the Paris agreement are also worried that offsets could be double-counted between the Paris standards and this agreement.[22] By some estimations, the ICAO standard leaves a gap between "carbon-neutral growth" and projected emissions of about 7.8 billion metric tons of carbon dioxide emissions.[23]

So what needs to actually happen? Airlines have a vested interest in reducing their fuel use and improving efficiency, since about one-third of their operating costs are spent on fuel, and they have done a lot to improve efficiency already, as we saw above. Though there is no international standard for sustainable fuel, some airlines and airports have started using alternative fuels—biofuels, agricultural waste, used cooking oil—to reduce their emissions, but these are either not available in large enough quantities or are too expensive.[24] Plus, on international flights, there are fights about who should get the credit for reducing the emissions, because we are all toddlers at heart. In *Drawdown*, the book of solutions I mentioned earlier, the authors found that wide-scale adoption of the most fuel-efficient airplanes, retrofitting some existing ones, and getting rid of older, inefficient ones could avoid 5.1 gigatons of carbon dioxide emissions and save $3.2 trillion on jet fuel by 2050.[25]

With airplanes, there is also stuff that we, the passive aeronauts, can do. One round-trip flight from New York to California generates about 0.9 metric tons of carbon dioxide per person, according to some estimates,[26] and in 2014 the average American's annual carbon footprint was about 16.4 metric tons,[27] so taking fewer flights can do a lot to shrink that. If it's a long flight, flying is probably more efficient than driving: cruising at altitude requires less fuel than other stages of flight. So flying from coast to coast would be better than driving solo, but if it's a short trip, it could be better to drive.[28] About 25 percent of airplane emissions come from landing and taking off, including taxiing, which is the largest source of emissions in the landing-takeoff cycle, so flying direct can help.[29] For all the bigwigs out there worried about their carbon footprint: fly coach. According to a study from the World Bank, the emissions associated with flying in business class are about three times as great as flying in coach.[30] Those bigger seats in business and first classes mean that fewer people are being moved by the same amount of fuel. The study estimates that the carbon footprint of a first-class seat could be as much as nine times bigger than an economy one. Some airlines are more efficient than others, too, either because of the planes they use or because they make efficient use of their space by carrying a lot of cargo. If you're flying across the Pacific Ocean, better to fly Hainan Airlines or All Nippon Airways (ANA), the most efficient airlines, and 64 percent more efficient than Qantas, the least efficient airline, according to the International Council on Clean Transportation. (Qantas, for what it's worth, disputes the study determining this ranking, which was performed by the same crew who exposed Volkswagen for gaming the emissions tests for its diesel cars. The head of fuel and environment for Qantas told the *Guardian* that the airline doesn't perform well by the terms of the study because it flies long distances, has premium cabins, and uses large aircraft, which is...the point.)[31] For the Atlantic, Norwegian is 51 percent more efficient than British Airways, though for both oceans, there is a big range. The International Council on Clean Transportation's rankings, available online, can help you out.[32]

You can offset your own travel, whatever the airlines are doing. When you buy carbon offsets, if done right, you're paying to take carbon dioxide out of the atmosphere in exchange for what you're causing to be pumped in. Some airlines offer offset purchases through their websites, but they don't make it easy—I have tried. You can buy them through third-party organizations, as well. I have used Sustainable Travel International, which also runs United Airline's offset program. For the trip I took in April 2017 to California for research on Silicon Valley, I generated 1.8 tons of carbon dioxide, which cost $47.31 to offset. (It's not cheap, but if I have enough money to take that trip, I should have to pay for the consequences, I think.) From there, I can choose whether this money goes toward reforestation projects or alternative energy development. You can use another organization, but make sure the program meets a few requirements: it should be verified by an independent third party, and the offsets should be additional (meaning that those trees weren't being planted anyway, or that wind turbine wasn't going up with or without your cold hard cash). There is some debate about the best way to offset—where and when trees should be planted—and if the cost of carbon dioxide emissions used by these organizations really captures the full long-term cost of greenhouse gas emissions writ large.

But for now, that's what we have. It's still not enough. For those who can afford it, it's great that there are several flights a day between New York and California, or New York and London, or New York and Tokyo. But that's also insane. There's a flight every hour between New York and Boston, and between New York and Washington. (There are also trains, might I add, which still result in greenhouse gas emissions, but much less.) But it maybe would also be okay if there were fewer flights, and if the cost of flying reflected the ultimate costs of pollution and climate change. We can't afford to emit all of those greenhouse gases from airplanes just because it's possible.

Not long ago, I spoke with Malte Humpert, a writer and researcher on the Arctic, who told me that before the end of the Cold War, very few, if any, planes flew over the North Pole because it was in Russian airspace. Now, there are about two to three hundred flights that pass over it, and they do so at a lower altitude, because otherwise it would be too cold for the planes.[33] Those planes, in addition to the carbon dioxide they emit, also emit particulate matter, which lands on the ice that they are now closer to. In the next chapter, you'll see why this is another long-term cost we truly can't afford.

Shipping: The World in a Box

Unlike in the international air cargo industry, when it comes to international shipping (on actual ships), not everything is coming up roses. (I mostly wanted to use that phrase, although we could have tried "when it comes to the environmental impact of the international shipping industry, it pays to take off your rose-colored glasses.")

Now that we've gotten that out of the way...If 1 percent of international trade by volume happens on a plane, almost all of the rest happens on a ship. All told, 90 percent (or 11 billion tons) of stuff—oil, clothing, corn, bananas, sneakers, speakers, absolutely anything—travels from one country to another on oceans, rivers, and seas. Almost everything you have that isn't from the US came to you on a ship.

Mostly, it all came to you on one type of ship, one that has most transformed the way that we produce and consume almost everything, and that is the container ship. You know a shipping container when you see one: a corrugated metal box—the empty, hulking, skeleton of global trade, resembling a block in a giant's Lego set.

Until 1956, goods were packed by dockworkers onto ships, piece by piece, or crate by crate, eventually filling up the holds of ocean liners that transported a limited amount of goods around the world. But, that year, Malcom McLean changed everything. McLean, a North Carolina truck company owner, was frustrated by how long it took for the goods he sent up to New York to be packed up and sent overseas, so he devised a plan: take the truck trailers themselves already loaded with goods and put them onto the decks of ships. In 1956, the *Ideal X*, his first cargo ship, set "sail," loaded with fifty-eight containers. The next year, he brought out a new ship, *Gateway City*, which could carry 226 containers, stacked one atop the other.[1]

McLean's invention started a revolution: ships could leave port more quickly, since the containers could be filled before they arrived and loaded right away; it drastically reduced the need for labor (meaning fewer jobs); and once the goods arrived at their destination ports, they could leave more quickly, loaded directly onto trucks or train chassis and sent trundling to where they were needed. Over time, the size of containers became standardized, and by the 1970s, the majority of consumer goods coming to the US were arriving in a container.[2] Now, ships have gotten bigger and bigger (and there are more and more of them) because larger boats meant that the price of sending each container went down. Now, about 120 million container loads travel across earth's waters each year, carrying goods worth about $4 trillion.[3] The bigger ships needed bigger ports to accommodate them, so ports all over the world have been dredged and rebuilt with bigger and bigger yards to fit all of those containers. For instance, 75 percent of mangroves, among the most powerful carbon-sequestering plants, in Shenzhen, China, disappeared from 2013 to 2016 for port expansion and land reclamation.[4] The Panama Canal has been widened so that these lumbering ironsides can make it through; some of the biggest ships are too big to fit in any American port. The biggest ship in the world can carry more than 20,000 containers at once.[5] We have officially lost control of the machine.[6]

Anyway, the ease of moving goods around and the rapidly declining price of doing so incentivized a decentralized manufacturing process (a lyrical phrase that conjures romantic visions, to be sure). There was no longer a real advantage to having an industry in one place, near a port, making obsolete the garment industry in New York, since, with free trade on the rise, cotton from the South could be shipped to Indonesia to be spun into fibers, which could be shipped to Bangladesh to be made into a T-shirt, which could then be sent back to the US to be sold—and the cost of shipping it all would be mere cents.[7] (Not to mention that manufacturing goods abroad often meant that companies could get away with paying workers much less and avoid environmental regulations, which is not a good alternative, but they do it anyway.) Half of the growth in emissions in China since 1990 comes from the offshoring and globalization of manufacturing industries. While the number of things manufactured in the US has declined over this period, we have become China's biggest customer.[8] Effectively, we've outsourced our emissions to China's factories, patting ourselves on the back as they become the world's biggest emitter.

That, as a coincidence that I did *not* see coming, is a good segue to talking about why shipping matters for the planet. While shipping industry white papers and representatives will tell you that shipping goods is the most energy efficient way to move things around (the least amount of fuel per item moved), that kind of misses the point. (And it is true only if ships are running with full cargo loads all the time, but there are currently too many ships, so a lot of ships are going back after dropping off their wares empty or partly empty, especially now that China isn't taking recyclables from the rest of the world.)[9]

Globally, the shipping industry accounts for about 3 percent of carbon dioxide emissions, and global trade is expected to keep growing.[10] At current rates of growth, by 2050, the shipping industry could be 2.5 times its current size and generate 17 percent of global greenhouse gas emissions.[11] Part of the reason the industry causes such a large portion of planet-warming gases is that these ships, for the most part, run on bunker fuel: number 6 fuel oil, the sticky mess that is left over from the process of petroleum refining, after all the lighter, cleaner fuel is distilled out—that is, the bottom of the barrel. The only thing heavier and dirtier than bunker fuel is the residue used to make asphalt. One description I came across said: "a person can walk on it when it's cool" and that it has the "consistency of peanut butter."[12] Except it's not peanut butter or something you could walk on—it's oil. This type of fuel can contain up to 1,800 times more air-polluting sulfur than even the diesel fuel burned in buses and big rig trucks. Ships use so much fuel that it is measured not in gallons but in metric tons per hour; big ships burn 200–400 tons per day when they are sailing.[13] (It had also been used until 2015 as home heating oil in New York City in 1 percent of the buildings, but accounted for about 85 percent of the city's soot emissions.)[14] While the emissions intensity of a given journey is decreasing overall, emissions from global shipping are increasing because ships are traveling farther, which more than offsets their gains in operational efficiency, and ships are also speeding up, which causes more fuel to be burned and therefore more emissions.[15]

What does it mean for fuel to be dirty, exactly? It means it produces more than just carbon dioxide: nitrogen oxides, which can lead to smog formation, and fine particulate matter, which cause lung cancer and increased risk of bladder cancer, according to the World Health Organization; sulfur dioxide, which also creates fine particulate matter and leads to ocean acidification; and black carbon, which is basically soot and also may be the second biggest contributor to climate change, after carbon dioxide.[16] (Black carbon, I'll deal with *you* later.) Bunker fuel can have as much as 3,500 times more sulfur oxides than road diesel,[17] releasing significant quantities of it into the air (by some measures, one container ship could release as much of this type of pollution in a day as about 500,000 big rig trucks.)[18] All of this pollution isn't just being spewed into the air over the open ocean either: along the main trade routes, 70 percent of emissions from ships occur within 250 miles of coastlines.[19] In Hong Kong, one of the world's ten dirtiest ports, between one-third and one-half of all the airborne pollutants come from ships.[20] But some of the pollution, obviously, affects the oceans: these pollutants have also been found to increase ocean acidification during summers in parts of the Northern Hemisphere with heavy ship traffic.[21]

In the US, some of these dangers have been recognized: in 2012, the EPA banned dirty fuel within two hundred miles of our coasts, requiring that ships use fuels with a sulfur content of 10,000 parts per million, falling to 1,000 ppm by 2015, and achieve an 80 percent reduction in nitrogen oxides by 2016. In its rationale for the new rule, the EPA estimated that using cleaner fuels could help avoid up to 14,000 premature deaths per year due to air pollution by 2020, and another 16,000 by 2030.[22] This has helped clean up the air we breathe on American coasts, especially in port cities. But some ships have found a way around this: they might wait outside the two-hundred-mile boundary until they have a spot in the port, burning more and more of their dirtiest fuel until the last possible second.[23] The international community has woken up a little, too. International shipping emissions were also not covered in the 2015 Paris accord, but in 2016, the International Maritime Organization, another UN agency, struck an agreement to reduce carbon emissions by at least 50 percent compared to 2008 levels by 2050, by improving efficiency at a faster rate, transitioning toward low- and zero-carbon fuels, optimizing trade routes, and using the full capacity of ships.[24] They have also set new limits on sulfur in fuel oil globally, which will take effect in 2020.[25]

There's also the question of what happens to the ships when they are no longer needed or have been replaced by something newer and bigger: they get scrapped. Guess what? Ship-scrapping is also heavily polluting. The EU has regulations about safe scrapping, but ship companies often change which country they are registered in so that they can scrap somewhere with fewer regulations, like India, Pakistan, or Bangladesh, where most scrapping takes place. In 2009, 40,000 mangroves were chopped down to accommodate a scrapyard in Bangladesh. Pollution from the scrapping over time has led to the extinction of twenty-one fish and crustacean species.[26]

To say that shipping goods is efficient really misses the point. Clearly, shipping, as currently practiced, is an incredibly damaging thing we do to our planet, and I haven't even talked about black carbon yet.

Black carbon, which primarily comes from those engines burning diesel-based bunker fuel, as well as cookstoves, wood burning, and forest fires, is the second biggest type of emissions from ships after carbon dioxide. Over a twenty-year period, black carbon can cause 3,200 times more warming than carbon dioxide.[27] Here is how: the particles that make up black carbon absorb sunlight, generating heat. Black carbon may stay in the atmosphere for only a few weeks at a time before falling out, so its effects are more regional than carbon dioxide, for instance.[28] This has profound effects for the polar ice caps. If black carbon happens to get on top of snow and ice, it absorbs heat, and also accelerates the melting process. Part of the reason that the melting of the polar ice caps matters for climate change is because they reflect so much sunlight (and therefore heat) to the atmosphere, instead of absorbing it. This is known as the albedo effect. If black carbon gets on the ice and snow, it diminishes the strength of the albedo effect, leading to more melting and therefore more warming. Black carbon may be responsible for more than 30 percent of the warming in the Arctic, which is warming at twice the rate of the rest of the world; it is also driving the melting of Himalayan glaciers and may also be having an effect on the reduction of the snowpack in the Pacific Northwest.[29] The Center for Climate and Energy Solutions estimates that black carbon is responsible for somewhere between 1.28°C and 1.68°C of warming in the Arctic, with 0.5–1.4°C of that warming taking place between 1976 and 2007.[30] Because of its particular warming effect on snow-covered regions, heavy fuel oil, which has the most black carbon, is banned from Antarctica both as a fuel and as cargo on oil tankers.[31] So far, it hasn't been banned in the Arctic. If heavy oil is spilled, it's much harder to clean up, especially in cold water, because it tends to sink or emulsify, or it sticks to sea ice, where it could further accelerate warming. Currently, heavy fuel oil is the most commonly used fuel in the Arctic, responsible for 68 percent of black carbon emissions there, mostly by fishing boats, cargo ships, and service vessels.[32]

That alone would be enough to worry about, but now let's get more worried, because what is this book for if not to encourage periodic panic attacks? Currently, there is relatively very little cargo shipped through the Arctic, but as ice melts (and decreases in thickness by about half, and summer ice disappears, likely around 2040),[33] routes mostly just imagined—directly over the North Pole, for instance—may open up to shipping and resource exploration: the Arctic is home to as much as a quarter of the world's known oil and gas reserves,[34] and there are lots of fish there, too. Some countries are banking on ice-free waters: Russia has opened up oil fields in Siberia and built ports along its northern coasts,[35] Arctic shipping could cut down the length of voyages between Europe and Asia to less than three weeks,[36] and China has ambitions to be a "polar great power."[37] Researchers anticipate that Arctic shipping could grow more than 50 percent between 2012 and 2050, causing 120 percent more carbon dioxide and black carbon emissions, which would, in turn, cause more rapid warming.[38]

This is all very scary. We are only beginning to understand how the presence of Arctic sea ice affects our weather patterns, though we know that its melting will change them. The good thing is that cutting down on black carbon emissions could make a big difference relatively quickly. According to a report from the UN, reducing black carbon and ozone in the lower part of the atmosphere (the kind we don't want), particularly in those countries closest to the Arctic, could cut warming in the Arctic by two-thirds by 2030.[39] If other measures were taken to reduce black carbon—forest fire reduction, clean cookstoves, cleaner industry—as well as reduce methane and ozone emissions, as much as 0.5°C of warming could be avoided,[40] which, when you're desperately trying to stay below 2°C of warming and we already have warmed by at least 0.85°C since pre-industrial times, is a sizable portion.

But doing that would be too easy. Let's explore a major way we've made this even more difficult: cars and trucks!

Cars, Trucks, and Justice

Given what you now know about boats and planes, it's clear that the way we're moving ourselves and our things around uses a lot of energy and causes a lot of pollution above and beyond carbon dioxide emissions, though the carbon dioxide emissions are also enormous. And we haven't even gotten to cars and trucks.

In 2016, emissions from transportation surpassed those from electricity generation for the first time in decades, making up about 28 percent of all the emissions produced in the US.[1] Of those, about 60 percent are from cars, and 23 percent are from trucks, together making up just under a quarter of all of our greenhouse gas emissions.[2] The huge amount of emissions associated with cars is why people often talk about the particular effectiveness of any emissions-saving solution in terms of the number of cars it would take off the road, because each one is a four-wheeled existential threat.

As we know from our look at e-commerce, a lot of our cargo also moves around by truck and more of it is moving more quickly. Sometimes, it makes the whole journey by truck; other times (for instance, international/transcontinental cargo), it travels by truck and something else. Most of the time, the way that something gets from warehouse/port/airport/railyard to you is by truck—in fact, about 70 percent of all domestic trade by weight happens on a truck, about $55 billion each day.[3] Heavy-duty trucks make up about 5 percent of all the vehicles on the roads, but they account for about 20 percent of all transportation emissions,[4] which leads me to believe (and also know for a fact) that they are pretty inefficient machines: government regulations stipulated that trucks get about seven miles per gallon in 2017. Seven! Miles! Per! Gallon! (Up from six! Miles! Per! Gallon! In 2010.)[5] These trucks run, almost exclusively, on diesel fuel, which is not as dirty as bunker fuel but emits more harmful compounds and particles than regular gasoline.

And then there are glider kits—a benign name for a really terrible thing. These are new trucks that are built without an engine so that older engines can be installed, thereby avoiding the more stringent efficiency regulations that apply to new engines. Glider trucks can be up to fifty-five times more polluting than regular old normally polluting ones and account for about 5 percent of all heavy-duty trucks currently on the road.[6] The Obama administration estimated that gliders alone could generate about one-third of the nitrogen oxide pollution from all trucks, which causes smog, acid rain, and many of the other health impacts we've talked about.[7] In regulations published in 2016, they limited the production of gliders to three hundred by the end of 2019, hoping to phase them out entirely.[8] Of course, Scott Pruitt's last action as EPA administrator was to tell manufacturers he wouldn't enforce the cap,[9] though this action was challenged in court and quickly undone by his replacement, Andrew Wheeler.[10] During his confirmation proceeding in early 2019, Wheeler informed Congress that he intends to retool the rule to protect glider manufacturers, so the policy is still up in the air.[11]

But enough about me and my glider kits—let's talk about you. Remember how you were ordering stuff online and a lot of it was being delivered by truck because you decided you needed it tomorrow? Remember that? Remember how great that was and how it meant that maybe a truck was traveling at less-than-full capacity because you needed your socks from cheapsocks.com? Research from the Environmental Defense Fund has found that 15–25 percent of trucks on the road are empty and, when they're not empty, they are 36 percent underutilized. If we used up half this currently unused capacity, we could cut emissions by 100 million tons each year, or about 20 percent of all freight truck emissions, and save $30 billion fuel.[12] Just saying.

And now we talk about cars. In 2014, we drove 344 million miles each hour, together, as one nation.[13] Over the last few years, emissions from vehicles have declined because of more efficient engines, but there has been a tiny uptick, largely because of the inexpensive price of gasoline.[14]

If you are traveling back in time and reading this in the summer of 2018, you have probably been hearing a lot about fuel economy standards in cars—how the Trump administration's plan to weaken them will have disastrous consequences in the fight to prevent some of the worst effects of climate change. If things continue on their current path, these lower fuel standards will cause the US to increase greenhouse gas emissions by an amount greater than what some countries emit in an entire year.

All of which is to say that fuel economy standards—combined with zero-emissions vehicles, like electric cars—can make a big difference. The Obama administration's plan would have required car companies to double the average mileage per gallon of cars, pickup trucks, and SUVs by 2025.[15] The Trump administration's plan would halt any efficiency improvements at the 2021 standards.[16] Always better to rest on one's laurels—isn't that the conventional wisdom?

Not to mention the fact that internal combustion engines, which cars use, are wildly inefficient to begin with. They waste between 70 and 88 percent of the gasoline we fill them up with.[17] And we have radically changed the way we organize space to accommodate the car: there are at least 3.4 and as many as eight parking spaces for every car in the country, enough to cover half of New Jersey;[18] nearly 18,000 square miles of land in the US is taken up by a road,[19] which is a little bit bigger than Maryland and a little smaller than West Virginia. Producing cement, the main ingredient in concrete, which is used in the construction of a lot of roads, accounts for about 5 percent of global carbon dioxide emissions and is the second-most-consumed substance on earth after water. Three tons of concrete per person are produced every year, each ton producing another ton of carbon dioxide emissions.[20]

All told, cars and trucks, if they become more efficient, or stop emitting, or we stop using them, have an enormous impact when it comes to climate change. But if I left it at that, you would be disappointed, so I actually want to talk about the disproportionate effects of air pollution from vehicles.

As we know from our previous study of jet fuel and bunker fuel, burning fossil fuels to power engines releases nitrogen oxides, which contribute to the formation of ozone in the troposphere and produce small particulate matter. Nitrogen dioxide, ozone, and PM 2.5 (the smallest size of this particulate pollutant) are three of the six criteria air pollutants, which are strictly regulated by the Environmental Protection Agency.[21] Nitrogen dioxide is known to cause asthma, low birth weight, and other respiratory problems.[22]

As is so often the case with pollution—we saw it with coal ash and nutrient pollution—air pollution affects different communities differently, which you might think sounds odd. In a city, you're saying to yourself, none of the air is clean, so it's all probably equally dirty. Not quite. A study published in 2017 found that while nitrogen dioxide decreased by at least one-third everywhere in the US from 2000 to 2010, it continued to harm nonwhite communities at a disproportionate rate. In 2000, nonwhite minorities were exposed to 40 percent more air pollution; in 2010, it was 37 percent. If people of color were exposed to the same amount of nitrogen dioxide as whites during those ten years, 5,000 premature deaths could have been avoided. In the US, racial minorities and low-income households are more likely to live near a major road (27 percent compared to 19 percent of the population as a whole), and air pollution is at its highest concentrations along major roads, where concentrations of nitrogen dioxide are on average 2.9 times higher than typical urban levels.[23]

Pollution, wherever it comes from, generally goes hand in hand with inequality. Differing effects of pollution are created by inequality, and pollution exacerbates unequal dynamics by creating an inequality of health outcomes, based solely on where we live, itself often a result of the historical legacy of racism. The disproportionate effects of air pollution in the US can be attributed to government-sanctioned discriminatory housing policies, which kept African Americans and other people of color from living in white neighborhoods, often ensuring that they would be near major roads or industrial facilities, which also spew pollution. Those policies further exacerbate economic inequality, which helps maintain that segregation. We already knew from existing studies that racial minorities and low-income households were exposed to industrial air pollution at greater rates and that the trend had persisted over time.[24] What we didn't know, and what these authors demonstrated, was that race as an isolated factor apart from income determined the level of exposure to these harmful pollutants.

It's much easier to conceive of industrial pollution than pollution coming from cars and trucks. Transportation pollution means that we're all responsible for it in a direct way, and that we're all exposed to it almost all the time, no matter where we live. But some of us suffer much more than others, and that's not right.

And it's not just a problem here. In Europe, the unintended consequences of policies aimed at reducing carbon dioxide emissions have resulted in disproportionate effects on non-white communities as well. Most countries in Europe encouraged and subsidized the use of diesel engines, because they release less carbon dioxide, but they emit more nitrogen dioxide, ozone, and particulate matter. (They also release much more nitrogen oxide pollution than the manufacturers told the public—see the Volkswagen emissions scandal of 2015.)

As a result, nitrogen dioxide pollution in European cities has shot up. In early 2017, London saw record levels of pollution, much of it from these diesel engines. According to the Department for Environment, Food, and Rural Affairs, nitrogen dioxide contributes to 23,500 deaths each year in Britain, the second-highest number in the European Union after Italy.[25] In a study looking at the sociological dimensions of air pollution in England and the Netherlands, scientists found dynamics similar to the United States. In both the Netherlands and England, neighborhoods where the non-white population made up more than 20 percent also had the most pollution. Neighborhoods in these countries had a significant correlation between poverty (a less strong predictor in the US) and exposure to pollution, because poor communities are often closer to highways or factories.[26]

Air pollution everywhere is a problem that requires our attention. It causes illness and premature death, and much of it could be alleviated by switching to mass transit, renewable energy, electric vehicles, and other solutions. But in our efforts to clean up our air, we should also focus on equality: narrowing the gap between those who have benefited from centuries of racial privilege and those who have suffered its effects. Time to see if ridesharing will help . . .

Hitching a Ride (share)

The year 2015 was a simpler time; the world's democracies didn't appear to be in danger; it was the hottest year on record (2016 would be hotter) but we didn't care! UberPool had arrived in New York, and damn if we weren't going to hop into a car with some strangers and possibly, as Uber reminds us in every corporate blog post about UberPool, meet the person we were going to marry.[1] (I did not do this.)[2]

What am I talking about? Your soulmate. That's what. But also Uber had announced its goal to take 1 million cars off the road in New York City by having more and more New Yorkers share car rides through UberPool,[3] which, as far as I can tell, is the type of Uber no one really wants to take.

In that post, Uber was confirming its professed aim of changing the way the world moves and, conveniently, announcing that that change will also make the world a better place. Classic Silicon Valley, am I right? We all get exactly where we want to go when we want to go there, possibly sharing a car with our soulmates, easing traffic, and preventing greenhouse gas emissions, too. A total dream.

It really is the typical Silicon Valley spiel, and some of it is probably sincere but still a little ridiculous: saving the world, while also generating billions of dollars of wealth, but who even cares because we are all part of the revolution, merely by tapping a few buttons on our phones and sharing all of our data.

Based on my observed experience in New York, I wasn't sure

about all of that. Yes, ride-hailing apps (which, after a regulatory decision in California, are known as transportation network companies,[4] so I am planning to abbreviate my references to ride-hailing companies as TNCs) provide the ultimate convenience, but as far as I could tell, traffic in New York, where I live, seemed to have gotten worse. I didn't know if this was because of TNCs, but it seemed like there was a possibility that their collective presence had added some cars to the roads.

So I wanted to know if I could believe Uber and its competitors, or if what I was seeing—more traffic and probably more cars on the road because of TNCs—was true. Was Uber living up to those early promises, or were we overlooking something important here? As we know, transportation and particularly cars are the biggest single source of emissions in the US, so whatever happens with cars has bigger implications, beyond the traffic in our own city or suburb.

TNCs are primarily used by young-ish people with disposable incomes, living in cities, which is not everyone, but the way the services they provide are used affect most of us, given that the majority of Americans live in cities or their greater metropolitan areas.[5] Let me just state that the problem is that America moves itself by car and not by public transportation or bicycles or walking, because so much of our built environment was built when the car existed, and so our society is largely organized around the car. Cars are the problem, but TNCs may be part of that problem. I also want to clarify the scope of what I'm talking about: I'm not talking about labor practices, corporate culture, efforts toward diversity and inclusion, data privacy and security, and taking advantage of protests against injustice to make money.

Within their first five years, from 2009 to 2014, TNCs managed to attract 250 million users around the world.[6] It took Uber six years to provide 1 billion trips;[7] the second billion came six months later.[8] That was in 2016. Since then, Uber has provided more than 10 billion trips globally.[9] In 2017, Americans took 2.6 billion trips with TNCs, a 37 percent increase over the previous year.[10] In 2018, trips taken with TNCs and taxis were expected to reach 4.74 billion, more than all of the trips taken by bus nationwide.[11] The question that logically follows is, how could these companies *not* be adding more cars to the road?

In most places, it would seem that TNCs are making things at least marginally worse, with some of the most annoying and bad things happening in New York. (A lot of this will focus on New York because New York City is one of the few cities that require TNCs to submit data about their activity.) But it doesn't have to be this way. Ergo, first, I will destroy TNCs and from their ashes I will build a new world order (of vehicular transportation).

If you live in a city, there's a pretty good chance that you've used a TNC yourself or gotten into a car summoned by a friend. According to one study that looked at seven major American metropolitan areas (New York, Los Angeles, Boston, Chicago, the Bay Area, Seattle, and Washington, DC), 21 percent of adults have gotten a car through an app themselves, and another 9 percent with a friend.[12] Across the country, about 60 percent of these trips would not have been taken if a ride-hailing service didn't exist, or they would have been taken by bike, foot, or public transportation. Twenty percent would have been in a personal car and the remaining 20 percent in a taxi.[13]

The argument that TNCs are more efficient for cities relies on the assumption that they are driving down personal car use—that fewer people are driving their own cars and are ridesharing instead. Some studies have suggested that the arrival of TNCs have caused some people to get rid of their cars or postpone buying one, but the replacement cars—the ride-hailing ones—are still adding driving miles to the road.[14] In New York, from 2013 to 2017, TNCs added 976 million miles of driving to city streets.[15] One study found that TNCs could be increasing the number of miles traveled by cars by 85 percent in part because of the distance a driver has to travel to pick up a passenger for a trip.[16] In the seven urban areas listed above, TNCs are likely adding miles, too; across the country, they have added 5.7 billion miles to our roads.[17] A lot of the time, the cars are driving around empty: by 2016, for about half of the miles they traveled in New York (600 million total), these cars had no passengers; in San Francisco, it was about 20 percent.[18] Meanwhile, car ownership has actually increased in nearly all American cities since 2012, and in most cases, grew faster than the population.[19] Plus, rates of car ownership among those who use public transit and TNCs are slightly higher than they are among those who only use public transit.[20] Either way, in New York City, there were 47,000 for-hire cars in 2013. By 2017, there were 103,000, 68,000 of which were affiliated with a ride-hailing app. Of those, all but 3,000 were affiliated with at least Uber.[21]

Another assumption is that more and more people will want to use the ridesharing services these companies offer, like UberPool. Lyft, another TNC service that offers shared rides (formerly known as Lyft Line), claimed that close to half of its trips in San Francisco and 30 percent of trips in New York were shared.[22] UberPool claims it provides 20 percent of trips in the places it's offered.[23] However, shared trips are not always given to those who ask: in New York, 37 percent of riders requested shared Lyft rides, according to the company, but only 22 percent of those people were matched with another passenger. In the same month, February 2018, UberPool saw similar match rates: 23 percent. Lyft's goal is for half of its rides to be shared in New York by 2022, which would still produce 2.2 miles driven for every mile of personal car driving taken off the roads.[24]

A third assumption, which is kind of like the first assumption though slightly different, is that TNCs are a complement to public transportation and not a substitute. That doesn't exactly bear out, either. That study I mentioned, which came out of the Institute for Transportation Studies at the University of California Davis, found that, overall, ride-hailing appeared to reduce public transportation use. TNCs generated a 6 percent decrease in bus ridership and a 3 percent decrease in light rail or subway ridership.[25] The New York subway system has a lot of problems, with or without the presence of TNCs, but the timing of this decline seems more than coincidental: in 2016, annual subway ridership fell by about 0.3 percent, to 1.756 billion trips per year, though weekday ridership was at its highest level since 1948; weekday bus ridership fell by about 1.6 percent that year, too.[26] Beyond New York, commuter rail, however, saw a 3 percent increase in usage, which makes sense, given that TNCs can give people a lift to or from their commuter rail station[27]—about 10 percent of commuters in Boston said they used TNCs to connect to transit on the way to work, and about 5 percent did so on the way home. The danger here is that some cities may begin to see TNCs as a way to fill in the gaps in their public transportation systems, without having to spend that much public money. City governments in Florida, Colorado, and New Jersey and transit agencies in Florida and California have provided subsidies to cover part of the fare for residents traveling to or from transit hubs. As long as these are extensions, and not replacements, for fixed bus or shuttle routes, they can be effective at boosting public transit usage, according to some studies. The fear is that the governments will cede these services to the TNCs, effectively stranding those who cannot afford even the subsidized rate or rely on public transportation more broadly.[28]

All of this adds traffic, which is both annoying and bad for us. In 2015, as a nation, we spent a collective 8 billion hours stuck in traffic, about 50 hours per driver.[29] All of that traffic is bad for air quality outside our cars—we know that because transportation pollution is regulated by the Clean Air Act—but it's also bad inside our cars. One study showed that pollution levels inside cars at red lights or in traffic jams are up to 40 percent higher than those outside, since the particles are more concentrated around your car than when traffic is moving and can come in through open windows or vents.[30]

Now, to rebuild from the ashes. But how? Well, the companies are doing a little. Lyft announced in spring 2018 that it would begin offsetting all of its trips, in order to be carbon-neutral.* Uber, which, like Lyft, doesn't own or lease the cars in its fleet (the drivers do), has very modestly begun encouraging drivers to switch to electric or hybrid vehicles with very modest sweeteners like an extra $1.00–1.50 credit per trip, with a maximum of $20 per week in some places; "educational assistance" in LA; and free charging in Sacramento.[31] You can also get an in-app notification if you're in an electric car, but, as of this writing, you can't request one.[32] Uber hoped to have 5 million electric or hybrid trips in 2018, up from 4 million in 2017 (data on the actual number is unavailable as of this writing).[33] Both Uber and Lyft have purchased or invested in bike-sharing or electric scooter–sharing companies.[34] (Uber is also adding electric tiltrotor aircraft.[35] Make of that what you will.)

But most TNCs are looking to the future, banking on widespread vehicle electrification over the next few years, as well as massive growth in automated vehicles, which Uber, Alphabet, and other companies are working hard on. Automated vehicles, in best-case scenarios, will make personal car ownership obsolete and will be more efficient—cars can platoon and reduce drag, don't need parking, and can offer ridesharing more effectively. One study, however, anticipates that automated vehicles could actually increase driving if vehicles are automated but not shared or electric: it will be easier to travel by car, so more people will do it; it will be cheaper since there won't be a driver; and it could

---

* Lyft's CEO said its offsets would be certified by 3Degrees, a consulting company that works with independent third-party organizations to verify their offsets, and would involve projects to reduce emissions through "automotive manufacturing process[es], renewable energy programs, forestry projects, and the capture of emissions from landfills."

further reduce public transportation usage, increasing greenhouse gas emissions by 50 percent. However, the study found, if automation, electrification, and ridesharing are combined, it could reduce emissions by 80 percent.[36] Let's do the second one.

There are some solutions that might help make city streets less congested and cause less pollution (some of which TNCs have resisted, arguing that they provide a service, especially to those who live in areas underserved by public transportation, which, okay, but that's not the vast majority of what they do). They include capping the number of vehicles, which New York has done (for a year, while it studies the issue), or capping the number of vehicles that can enter certain parts of cities.[37] Congestion pricing is another option, as is a tax on fares that would go toward public transportation or building electric vehicle charging stations.[38] (As of January 2019, in New York City taxis, a $2.50 surcharge on the fare for trips that go below Ninety-Sixth Street in Manhattan goes to the city's transportation authority, compared to a $2.75 surcharge for TNC rides, as part of a congestion pricing plan.[39] In March 2019, New York passed a congestion pricing plan that will begin in 2021. NYC is the first American city to enact such a plan, which have existed in other cities for years.)[40]

But having an efficient, low-carbon city or transportation network doesn't just happen, and I think it would probably be a mistake to assume that it will, just because we have apps that make it easier to get where you want to go. In the end, the problem isn't necessarily the apps; the problem is cars and that we all want to ride in them, almost all the time.

The problem is we want everything to be everywhere, and we want to be there too, as quickly as possible. We want to be cool in our homes without paying too much for it, get our things on demand, stream video, get in our cars or rideshares and do whatever else we want, and also have our southern forests protected. So really, once again, the problem is us.

Conclusion

In November 2018, I went on a reporting trip to Portland, Maine, for an article I was working on. The story was about how some people in Maine—businesspeople, elected officials, civil servants, etc.—were trying to make connections between Maine and Arctic countries, both as a present-day business venture (for trade between Maine and Norway, for example) and a long-term play for the opportunities that may come with climate change: when Arctic sea ice has melted, when Greenland is no longer a frozen tundra, when this last spot of wilderness is open to greater human involvement. That is quite literally another story, but while I was there, I went out to visit Peter Rand, an independent oyster farmer, who grows oysters on several small floating docks where the New Meadows River meets Casco Bay. His grandfather had built a small hunting and fishing cabin on the riverbank, and Rand had now started a small aquaculture business there. That day in November was bright and cold and crisp, and Rand was leaving the oysters to hibernate for the winter. But I got to try one that had been harvested that morning—it tasted like ocean in the best possible way. As we stood on the docks and I looked down at the clean, clear water and the fishing boats passed up and down the river heading out to sea, I felt momentarily connected to a world far from my own. Then, Rand told me that there were more and more oyster farmers in Maine, many of them former commercial fishermen, who could no longer support themselves with their ocean catch. Cod fishing in the Gulf of Maine has been strictly regulated for years, but the cod population still hasn't recovered from overfishing, and climate change is making it more difficult; the shrimp fishery was closed in 2014 and will stay closed through at least 2021; the herring fishery is on the verge of collapse.[1] Any illusions I had been quietly harboring about a world unspoiled quickly disappeared. Even here, in a place where people depend on nature, where a Maine oyster has value because of the "pristine" environment in which it is raised, nothing can be separated from the unconscious exploitation and damage happening elsewhere. It made me think again that there is no corner of the earth that our actions don't affect, and that our indifference to the value of our resources and to their limits threatens everything, including our own success and survival.

The Maine oyster farming industry is perhaps not the most pressing issue we have to deal with, but I think it speaks to the broader problem that I've tried to address in the book: we have pushed the planet to its limits, unconsciously sacrificing the future to meet the needs, real or imagined, of the present. In the name of convenience or immediate gratification or profit, we've created a world where we use resources because we can, with little attention paid to our waste and the problems it creates. We've imagined that our actions are not connected to each other, as if we don't live on one planet with one ocean and one atmosphere, one Arctic and one Antarctic. The crisis of climate change and environmental degradation feels sudden, but it's actually an escalation of urgency, building since the Industrial Revolution and known about since the 1970s. Our perceptions that our actions aren't connected and that the planet's abundance is unlimited and the behavior those perceptions encourage are catching up with us.

If you've gotten to the end of this book and you're upset, I apologize. But not really, because it's not my fault. If you're upset, that probably means I did something right, unfortunately. Reading about climate change and pollution and the hidden effects of our actions is really hard to confront and deeply upsetting. You may feel like I've laid out a set of enormous problems and not given you a way to solve them. But I don't think that's true. Yes, I have not given you a to-do list for fixing climate change—I didn't tell you to turn out the lights when you leave a room and bring a reusable bag to the grocery store. Those are good things to do, sure, because using fewer resources and creating less waste are good. But the problem, as we have seen, is much bigger than that. Instead, what I hope I've given you is enough information to see the shape of that big problem, and the context to understand what is needed to solve it.

But if you need a list, here are some major ideas that I hope the book has gotten across:

- Our actions and the problems they create are connected, all around the world. Goats in the Mongolian desert add to air pollution in California; throwing away a computer helps create an illegal economy that makes people sick in Ghana; a loophole in a treaty contributes to deforestation in the American South to generate electricity in England; our idea of the perfect carrot could mean that many others rot in the fields. We can't pretend anymore that the things we do and wear and eat and use exist only for us, that they don't have a wider impact beyond our individual lives, which also means that we're all in this together.

- A lack of transparency on the part of governments and corporations has meant that our actions have consequences we are unaware of (see above), and if we knew about them, we would be surprised and angry. (Now, maybe, you are.)

- It's important to understand your actions and larger social, cultural, industrial, and economic processes in context, because then you can better understand which specific policies and practices would make a difference, and what they would achieve.

- Living in a way that honors your values is important, even if your personal habits aren't going to fix everything. We need to remember what is at stake, and the small sacrifices we make may help us do that, if you need reminding. If we know what our sacrifices mean and why they might matter, we might be more willing to make them.

- When you hear politicians and business leaders and people in your own life say that transitioning to a green economy or building resilient infrastructure (to protect us from the coming disasters that humanity has created) will be too expensive, ask them these questions:

  1. Won't it be more expensive to lose cities to sea level rise and fight forest fires and deal with refugee crises spurred by drought and famine and other disasters, for instance?

  2. Couldn't mitigating climate change be a way to make money—by building new infrastructure, developing new technologies, providing new services, and avoiding chronic health costs, such as asthma, which can be caused by fossil fuel pollution?

3. Why wouldn't you just try, in case (almost) all the scientists are right?
4. Does it matter how much anything costs if you're dead and Miami is underwater?

- Those directly responsible for pollution have rarely, if ever, paid for the costs. And they've created a lot of pollution: greenhouse gas emissions, oil spills, toxic waste, every other kind of waste, chemical pollution, and on and on forever. We're the ones who have paid for it, and we'll continue to pay if we don't demand that they change their bad behavior. We suffer the effects of bad air and dirty water. We pay for the cost of the illnesses associated with exposure to toxic pollution. Especially in the United States, there is no reason why we shouldn't have access to clean water, or why people are choked by asthma and lung cancer because of where they live. We pay for the ultimate costs, too: sea level rise, stronger hurricanes, more punishing droughts, more frequent forest fires. But the low-level background pollution we all live with is part of the same problem: corporations aren't held accountable for the waste they create.

- It is always better to know more and make more informed decisions, but it should not be up to the consumer to figure out what is the most responsible or sustainable option. I shouldn't have to find out which brand of jeans uses the least amount of water; the brands should be more transparent and tell us how much water they use, and then use much less. They are the ones who are making money from our choices, and we should not support those who don't at least tell us what they're doing.

- If you care about justice, you care about climate change and pollution. If you want to fight inequality, you have to fight climate change, too. Already, the consequences of climate change disproportionately affect low-income communities and people of color in this country. That's largely because of centuries of injustice that have kept wealth from these communities, and directly or indirectly prevented them from living in less perilous places. Around the world, people living in poverty suffer most from climate change. Poorer countries are often in places with limited natural resources to begin with—like water or fertile land—or where the wealth has been taken out by Western colonizers and corporate interests. Or they are extreme environments to begin with: deserts, coasts, or islands far out in the middle of the ocean. People without means might be less easily able to recover from storms and fires, or they might not be able to find a better life if famine strikes where they live. The countries and communities that have contributed least to climate change and pollution will be the most affected. That is an injustice.

- We have to vote. Very few politicians (if any) have paid any price for not only not leading, but also for holding us back from addressing this impending disaster. They've made the prospects for fighting climate change worse, and they've made sure the effects of climate change will be more destructive and happen sooner. We can't allow that anymore. It's not enough to elect politicians who profess to understand and care about the problem; we must hold them accountable for their actions. Vote for people who offer meaningful policies, and make sure they are achieved. If they don't, vote for someone else who will.
- It's okay to be angry and upset, but you have to stay engaged and involved.
- Ask questions. Demand change.

Thank you for helping.

Afterword

Now that you've really read this whole book and gotten all the way to the second ending and are still reading, you clearly want something more. You've learned and hopefully internalized the lesson that collective action is what it takes to make progress on climate change. You are probably looking for more specific ways to do this. I hope, by this point, you aren't looking for me to tell you to stop eating red meat and ordering stuff online, because I'm just not going to do it!

What I can do is tell you that if you want to do something, one of the most effective things you can do besides voting is to talk about climate change. Only about a third of Americans say they ever talk about climate change. Once people do talk about it, though, they are more likely to consider climate change a risk and to support climate action as government policy.[1]

Another thing you can do is to join an environmental organization, preferably one in your community. I spoke to Kimberly Ong, a lawyer for the Natural Resources Defense Council, one of the country's largest environmental advocacy groups. She helped explain why supporting a national organization like the NRDC is a good thing to do but getting involved with a local organization is even better.

A lot of climate policy that actually makes a difference happens at the local level. Mayors and governors and state officials make decisions about where your electricity comes from, what kind of public transportation you have (or don't), whether a new road or highway gets built and where, if a new factory or polluting facility (like a fossil-fuel-fired power plant or an Amazon fulfillment center) comes to town, what kinds of regulations are on the books to stop pollution, or whether to move the state away from a reliance on fossil fuels.

National and international issues get all of the attention—the Paris Agreement comes to mind—and those things are really important. But neither the president nor your senator decides whether your state gets its electricity from natural gas or renewable energy—that happens in state government. This is another reason why it matters that you vote in *every* election, not just in the presidential elections every four years.

Generally speaking, it's harder to find out about environmental issues happening close to home than about big national issues that get a lot more media attention, like the Keystone XL pipeline. State environmental protection agencies have newsletters or bulletins where they publish everything that's happening—new projects, laws, or regulations—because they're required by law to do so. But these can be pretty boring, long, and complicated.

If you join a local environmental advocacy organization, they will have already sifted through all of that information and can tell you which projects or efforts most need your help and where you'll have the most power. You're the expert on your neighborhood and community: if a new power plant is being planned next to the part of the river where you launch your kayak or go for hikes, you'll know exactly what the consequences will be. If the facility is an especially polluting one, your health or that of your family might be affected.

However, communities of color, particularly Black communities, are disproportionately affected by pollution, and it is important to keep in mind that if some kind of fossil fuel power plant (for example, though there are lots of other sources of pollution) isn't going in your community, unless you shut it down for good it might have to go to someone else's (until we get rid of fossil fuels). No one should live with this kind of pollution, but those communities that have been more affected over time should not continue to be harmed. You should fight against the siting of pollution in your community, but if your community is less impacted, you should work with historically impacted communities in your state or local area to ensure that pollution doesn't continue to affect them, either.

You're also responsible for whether the politicians or other officials in decision-making positions keep their jobs if they make bad decisions (or good ones!), so they should listen to you. If they don't, you don't have to reelect them! You could even run for office yourself.

When Carl Shoupe, a retired and disabled coal miner living in Benham, Kentucky, got involved in local politics, it was because of the most local of issues: his electricity bill. He got himself elected to the city council to figure out where his money was going.

In the fifteen or so years since, through that work and by joining a local environmental organization, Kentuckians for the Commonwealth, Shoupe has become an outspoken advocate against fossil fuels and for a just transition. Working with the organization and its other members, he has helped stop mountaintop removal mining in his community (at least temporarily). The organization has also helped members of the community retrofit their inefficient homes, saving electricity, energy, and money.

"When you join an organization and you've got a question or you need something, the answer is right there," Shoupe told me. "If you're interested in your community and trying to help, it's imperative that you join a group that supports that issue that you're interested in."

"I came from a great family, a union family, and that's where I got all of this from," he said. "I was raised that way. By God, if something ain't right, speak up. It ruffles some people's feathers, but if you get enough people behind you, you can get things done."

Kimberly Ong said the same thing, more or less. "It's always good to have a buddy, but it's especially good to have a buddy in environmental advocacy."

Looking for something else? Try getting involved in a citizen science project. Since the 1960s, the Cornell Lab of Ornithology has been encouraging normal people to get out in their backyards and neighborhoods and count the birds that they see, and then share their results with the scientists at the lab.

Without the help of ordinary people looking for and counting birds where they live, the scientists wouldn't be able to gather the data they need to do their work and measure bird populations across thousands of miles or across many decades.

And that data is really, critically important. In 2019, the lab published a landmark study, showing that the bird population in the US and Canada had declined by about 3 billion individuals since 1970.[2] There are lots of reasons for that decline—the effect and persistence of chemicals in the environment, like the one that Rachel Carson identified in *Silent Spring*, for one. Another is habitat loss and fragmentation, which I also wrote about in the context of the COVID-19 pandemic and other zoonotic diseases in the foreword to this edition. Climate change makes all of these underlying conditions worse.

With the threat of climate change to animals and plants all over the world, the focus has been on extinction—what will be permanently lost by the increasingly variable and extreme conditions that climate change will bring. But what this study showed is that the overall decline in biomass—the quantity of living things—is also a serious problem with significant effects. As the study's authors wrote, "Declines in abundance can degrade ecosystem integrity, reducing vital ecological, evolutionary, economic, and social services that organisms provide to their environment. Given the current pace of global environmental change, quantifying change in species abundances is essential to assess ecosystem impacts."

If scientists don't know what is happening to bird populations (and those of other species), they can't understand how things are changing. They can't do any of that without help.

It will also be better for you, personally. Forget about the planet! (As if!) People who participate in these kinds of activities have been shown to be healthier and more physically active, less likely to experience depression, and feel a greater sense of purpose, according to several recent studies.[3]

Whatever you do, I would recommend starting with just doing one thing. It may not feel like enough, but everyone has to start somewhere. And we can't all do everything. We can't all save the wetlands, call our representatives, lead zero-waste lifestyles, ride our bikes to work, protest a new landfill, reform the recycling industry, invent fully biodegradable plastic, capture carbon, come up with a new battery for electric vehicles, and discover a completely renewable electricity source. But maybe we could each do one of those things, or something else entirely. We could participate in a citizen science effort or run for office on a platform of climate action and racial equity or join a local environmental organization. Once we do one thing, maybe we'll think, *That wasn't so hard; maybe there's something else I can do.* Or we'll keep doing that thing and make a difference by just the act of doing it and talking about it with our friends, families, neighbors, and colleagues.

But sometimes it helps to know that there are other people

doing really cool and important things. You've just read about two: Carl Shoupe and Kimberly Ong. They do are doing the hard and crucial work of making change on the local level, of holding government accountable—by running for office, working in state government, collaborating with local organizations, or using the law to hold polluters to account to make sure that our regulations protect all of us and to fight for the planet.

Now, you go try! We need everyone. This is our chance to change the world.

Acknowledgments

It feels wildly inadequate to merely acknowledge all of the people who helped me in writing this book, so I am acknowledging, instead, my profound gratitude to:

All of the climate scientists, environmental lawyers, researchers, activists, and advocates, without whom the future of the planet would be much worse off, who were so generous with their time and knowledge. I am astounded every day by their brilliance, their commitment to solving the problem, and their willingness to keep trying in the face of apathy and ignorance.

To my parents, whose extraordinariness I have trouble describing, for teaching me how to think and making me feel that what I think matters. I am so proud and so lucky to be your daughter, and I hope I can be like you when I grow up.

My sister and brother, Rose and Jack, for being the funniest people worldwide, and for always trying to teach me how to be cooler and better, which I am not.

Gretchen Young, my editor, for her enthusiasm, generosity of spirit, unflagging interest in this subject, and, above all, for making this book so much better with her masterful editing, careful eye, patience, and humor.

Esther Newberg and Zoe Sandler, my agents, for believing in this project from the beginning. Thank you, Esther, for taking me seriously. Thank you, Zoe, in general and for answering emails not just in the subject line like Esther.

Sean Lavery, for being the most incredible fact-checker, who kept me honest and accurate, and for going above and beyond that job description.

Everyone at Grand Central, especially Jimmy Franco, Amanda Pritzker, Emily Rosman, and Alli Rosenthal, for their hard work and creativity in helping this book succeed.

Henry Fountain, my friend and human calculator/science textbook, for improving the integrity and overall quality of the book.

John Broder, my editor for all time, the most encouraging and selfless friend and mentor I could have ever dreamed of.

All of my friends who helped me in one way or another, in particular Lilly Corning, Ariel Doctoroff, Hannah Goldfield, Lepi Jha, Margaret Katcher, Rory McAuliffe, Kathryn Olivarius, Justina Ray, Nick Roth, and Louisa Strauss. I'm sorry for only talking about coal ash for the last two years.

Mary, Garrett, Caleb, and Elizabeth Moran, for their general enthusiasm and unfailing generosity, love and support.

All of the people along the way who taught me about these issues and how they affect the lives of real people, including Beth Alexander, Thomas Cmar, Lisa Evans, Anne Davis, Linda Greer, Amanda Hawes, Anthony Leiserowitz, Peter Rogers, and Ted Smith.

All of the people who taught me how to be a reporter, by example, by editing my work, and by their friendship: Jan Benzel, Nina Bernstein, Adam Bryant, Henri Cauvin, Laura Chang, Annie Correal, Mary Duenwald, Celia Dugger, Janet Elder, Hannah Fairfield, Justin Gillis, Erica Goode, Elisabeth Goodridge, Marty

Gottlieb, Denise Grady, Michael Luo, Andy Newman, Matt Richtel, John Schwartz, Dan Sforza, David Shipley, Hiroko Tabuchi, and Ian Trontz.

All of my favorite writers and thinkers who have indelibly shaped my sense of the world and my place in it (most of whom are historians because old habits die hard, none of whom were directly involved in this book, many of whom I don't know personally, and all of whom may be surprised to be listed here if they even ever read this) and whose influence is all over this book: Thomas Andrews, James Baldwin, David Blight, W. Jeffrey Bolster, William Cronon, John Mack Faragher, Paul Grant-Costa, Pekka Hämäläinen, Nikole Hannah-Jones, Elizabeth Kolbert, Jill Lepore, Sarah Lyall, Cindylisa Muniz, Barack Obama, Marci Shore, Fred Strebeigh, and Richard White.

Finally, and also most important, George, my husband, whom I met in an UberPool. The most fun world to imagine is the one that we get to live in together, so I hope it spins on as a habitable place for human beings forever. Thank you for believing in me and in the importance of my work even though you literally spend your days saving lives.

Notes

Preface

1. Karn Vohra, Alina Vodonos, Joel Schwartz, et al., "Global Mortality from Outdoor Fine Particle Pollution Generated by Fossil Fuel Combustion: Results from GEOS-Chem," *Environmental Research* 195 (April 2021).
2. Christopher W. Tessum, David A. Paolella, Sarah E. Chambliss, et al., "$PM_{2.5}$ Polluters Disproportionately and Systemically Affect People of Color in the United States," *Science Advances*, 7, no. 18 (April 28, 2021).
3. Xiaodan Zhou, Kevin Josey, Leila Kamareddine, et al., "Excess of COVID-19 Cases and Deaths due to Fine Particulate Matter Exposure During the 2020 Wildfires in the United States," *Science Advances*, online, August 13, 2021.

Introduction

1. EPA, "Coal Ash Basics," https://www.epa.gov/coalash/coal-ash-basics. In 2014, 130 million tons of coal ash were generated in the US.
2. World Wildlife Fund, "Living Waters: Conserving the Source of Life," 2002, p. 10; Deb Berlin, "My Jeans Are Very Thirsty!" EPA Blog, June 14, 2010, https://blog .epa.gov/2010/06/14/my-jeans-are-very-thirsty.
3. Livia Albeck-Ripka, "How to Reduce Your Carbon Footprint," *New York Times*, https://www.nytimes.com/guides/year-of-living-better/how-to-reduce-your -carbon-footprint.
4. Nathaly Gonzalez, Euromonitor, email to author, February 28, 2019.
5. Susan Gonzalez, "'I'm Optimistic Because of You,' Former Secretary of State John Kerry Tells Students," *Yale News*, April 28, 2017, https://news.yale .edu/2017/04/28/i-m-optimistic-because-you-former-secretary-state-john-kerry -tells-students.
6. David Roberts, "Alexandria Ocasio-Cortez Is Already Pressuring Nancy Pelosi on Climate Change," Vox, November 15, 2018, https://www.vox.com/energy-and

-environment/2018/11/14/18094452/alexandria-ocasio-cortez-nancy-pelosi
-protest-climate-change-2020.

Technology and the Internet

1. Lotfi Belkhir and Ahmed Elmeligi, "Assessing ICT global emissions footprint: Trends to 2040 & recommendations," *Journal of Cleaner Production* 127 (January 2018): 448–463.
2. Nicola Jones, "How to Stop Data Centres from Gobbling Up the World's Electricity," *Nature* 561 (September 13, 2018): 163–164.
3. Alan Meier, phone interview with author, April 1, 2016.
4. United Nations Environment Programme, *Waste Crimes, Waste Risks: Gaps in Meeting the Global Waste Challenges* (Nairobi, Kenya: United Nations Environment Programme, 2015), 7.

The Physical Internet

1. "Preliminary Monthly Climate Data," National Weather Service, accessed January 4, 2018, https://w2.weather.gov/climate/index.php?wfo=gjt.
2. John C. Fyfe, Chris Derksen, Lawrence Mudryk, et al., "Large Near-Term Projected Snowpack Loss over the Western United States," *Nature Communications* 8, no. 14996 (April 18, 2017): 1.
3. Philip W. Mote, Sihan Li, Dennis P. Lettenmaier, Mu Xiao, and Ruth Engel, "Dramatic Declines in Snowpack in the Western US," *NPJ Climate and Atmospheric Science* 1, no. 2 (2018): 4.
4. "Qwest Completes Purchase of US West," *New York Times*, July 3, 2000, https://www.nytimes.com/2000/07/03/business/qwest-completes-purchase-of-u-s-west.html.
5. "CenturyLink and Qwest Complete Merger," CenturyLink, April 1, 2011, http://news.centurylink.com/2011-04-01-CenturyLink-and-Qwest-Complete-Merger.
6. Tomas Nonnemacher, "History of the U.S. Telegraph Industry," https://eh.net/encyclopedia/history-of-the-u-s-telegraph-industry.
7. Stephen A. Ambrose, *Nothing Like It in the World: The Men Who Built the Transcontinental Railroad, 1863–1869* (New York: Simon and Schuster, 2000), 24.
8. Ingrid Burrington, "How Railroad History Shaped Internet History," *The Atlantic*, November 24, 2015, https://www.theatlantic.com/technology/archive/2015/11/how-railroad-history-shaped-Internet-history/417414.
9. Brandon Yergey, CenturyLink, email to author, January 6, 2018.
10. Jane Tanner, "New Life for Old Railroads; What Better Place to Lay Miles of Fiber Optic Cable," *New York Times*, May 6, 2000.
11. Ingrid Burrington, "Why Are There So Many Data Centers in Iowa?" *The Atlantic*, December 1, 2015, https://www.theatlantic.com/technology/archive/2015/12/why-are-so-many-data-centers-built-in-iowa/418005/

12. Paul E. Ceruzzi, *Internet Alley: High Technology in Tysons Corner, 1945–2005* (Cambridge, MA: MIT Press, 2011), 1–17.
13. Ceruzzi, p. 154.
14. Ingrid Burrington, "Why Amazon's Data Centers Are Hidden in Spy Country," *The Atlantic*, January 8, 2016, https://www.theatlantic.com/technology/archive/2016/01/amazon-web-services-data-center/423147.
15. "Data Centers," Loudoun County Economic Development, https://biz.loudoun.gov/key-business-sectors/data-centers.
16. Kate Royce, "Apple's First Retail Store Seeks New Core Buyers in Washington, D.C. Area," *Washington Times*, May 16, 2001.
17. "Apple Store Opening Draws Rave Reviews," CNET, January 2, 2002, https://www.cnet.com/news/apple-store-opening-draws-rave-reviews.
18. International Energy Agency, "Digitalization and Energy" (Paris: International Energy Agency, 2017): 21; SINTEF, "Big Data, for better or worse: 90% of world's data generated over last two years," May 22, 2013, https://www.sciencedaily.com/releases/2013/05/130522085217.html; Nicolaus Henke, Ari Libarikian, and Bill Wiseman, "Straight talk about big data," McKinsey Quarterly, October 2016, https://www.mckinsey.com/business-functions/digital-mckinsey/our-insights/straight-talk-about-big data; Bernard Marr, "How Much Data Do We Create Every Day? The Mind-Blowing Stats Everyone Should Read," *Forbes*, May 21, 2018.
19. Cisco Systems, "Hyperconnectivity and the Approaching Zettabyte Era" (San Jose, CA: Cisco Systems, June 2, 2010), 2.
20. Cisco Systems, "The Zettabyte Era: Trends and Analysis" (San Jose, CA: Cisco Systems, June 7, 2017), 3.
21. Pew Research Center, "Internet/Broadband Fact Sheet," February 5, 2018, http://www.pewInternet.org/fact-sheet/Internet-broadband.
22. Pew Research Center, "Mobile Fact Sheet," February 5, 2018, http://www.pewInternet.org/fact-sheet/mobile.
23. Statista, "Percentage of All Global Web Pages Served to Mobile Phones from 2009 to 2018," https://www.statista.com/statistics/241462/global-mobile-phone-website-traffic-share.
24. Statista, "Mobile Phone Internet User Penetration Worldwide from 2014 to 2019," https://www.statista.com/statistics/284202/mobile-phone-Internet-user-penetration-worldwide.

Bringing the Cloud to Earth

1. Adam Nethersole, interview with author, January 17, 2018.
2. "Powering a Google Search," Official Google Blog, January 11, 2009, https://googleblog.blogspot.com/2009/01/powering-google-search.html.
3. Klint Finley, "The Average Webpage Is Now the Size of the Original *Doom*," *Wired*, April 23, 2016.

4. Sandvine, "Global Internet Phenomena Report," (San Jose, CA: Sandvine, 2018): 6.

5. Arman Shehabi, Ben Walker, and Eric Masanet, "The energy and greenhouse-gas implications of Internet video streaming in the United States," *Environmental Research Letters* 9, no. 054007 (2014):1.

6. Shehabi, Walker, and Masanet.

7. Sandvine, "Global Internet Phenomena Report," (San Jose, CA: Sandvine, 2018): 9.

8. Sandvine, "Global Internet Phenomena Report" (San Jose, CA: Sandvine, 2018), 7.

9. eMarketer, "US Digital Video Viewers, 2018–2022 (Millions and % Change)," February 22, 2018, https://www.emarketer.com/Chart/US-Digital-Video-Viewers-2018-2022-millions-change/216673.

10. eMarketer, "Average Time Spent per Day with Video by US Adults, by Device, 2015–2019 (Hrs:Mins)," September 29, 2017, https://www.emarketer.com/Chart/Average-Time-Spent-per-Day-with-Video-by-US-Adults-by-Device-2015-2019-hrsmins/211432.

11. eMarketer, "Average Time Spent per Day…"

12. Shehabi, Walker, and Masanet, p. 6.

13. NPD Group, "Entertainment Trends in America" (Port Washington, NY: NPD Group, 2017).

14. United States Environmental Protection Agency, "Overview of Greenhouse Gases," https://www.epa.gov/ghgemissions/overview-greenhouse-gases. Updated October 31, 2018.

15. Ashley Rodriguez, "Even with Streaming Video, a Third of Americans Still Buy and Rent," *Quartz*, November 24, 2017, https://qz.com/1136150/even-with-streaming-video-a-third-of-americans-still-buy-and-rent.

16. International Energy Agency, "Digitalization and Energy" (Paris: International Energy Agency, 2017), 105.

17. Eric Masanet, interview with author, December 1, 2017.

18. Uptime Institute, "Uptime Institute Global Data Center Survey" (New York: Uptime Institute, 2018), 5; "Efficiency: How We Do It," Google Data Centers, https://www.google.com/about/datacenters/efficiency/internal.

19. Natural Resources Defense Council, "Data Center Efficiency Assessment," 2014, p. 5

20. Nethersole.

21. "Hamina, Finland," https://www.google.com/about/datacenters/inside/locations/hamina.

22. "Odense Data Center," https://www.facebook.com/OdenseDataCenter.

23. "Luleå Data Center," https://www.facebook.com/LuleaDataCenter.

24. "What Is Bitcoin?" Coindesk.com, updated January 26, 2018, https://www.coindesk.com/information/what-is-bitcoin; Coindesk, "How Bitcoin Mining Works," updated January 29, 2018, https://www.coindesk.com/information/how-bitcoin-mining-works.

25. Jon Choi, interview with author, April 27, 2018.

26. Rachel Rose O'Leary, "A Multi-Million Dollar Bet Ethereum's Proof-of-Stake Isn't Coming Soon," Coindesk, October 4, 2018, https://www.coindesk.com/

a-multi-million-dollar-bet-ethereums-proof-of-stake-isnt-coming-soon; Yezi Peng, "Will Ethereum Be the Platform That Successfully Brings Blockchain into the Mainstream?" Harvard Business School Digital Initiative, https://digital.hbs.edu/platforms-crowds/will-ethereum-platform-successfully-brings-blockchain-mainstream.

27. Julianne Harm, Josh Obregon, and Josh Stubbendick, "Ethereum vs. Bitcoin," Creighton University, https://www.economist.com/sites/default/files/creighton_university_kraken_case_study.pdf; "Top 100 Cryptocurrencies by Market Capitalization," CoinMarketCap, accessed December 17, 2018, https://coinmarketcap.com.

28. Alex De Vries, "Bitcoin's Growing Energy Problem." *Joule* 2 (2018): 801–809.

29. "Bitcoin Network Average Energy Consumption per Transaction Compared to VISA as of 2018 (in Kilowatt-Hours)," Statista, https://www.statista.com/statistics/881541/bitcoin-energy-consumption-transaction-comparison-visa.

30. Joe Romm, "Bitcoin Energy Consumption Has Been More Overhyped than Bitcoin Itself," *ThinkProgress*, May 16, 2018, https://thinkprogress.org/bitcoin-energy-consumption-overhyped-56e018e7a6d8.

31. Tanwen Dawn-Hiscox, "Google Signs 10 year PPA with Netherlands' Largest Solar Farm," *DataCenterDynamics*, July 10, 2017, https://www.datacenterdynamics.com/news/google-signs-10-year-ppa-with-netherlands-largest-solar-farm.

32. Lee Bell, "One Third of Internet Users Visit an Amazon Web Services Cloud Site Daily," *The Inquirer*, April 19, 2012, https://www.theinquirer.net/inquirer/news/2169057/amazon-web-services-accounts-web-site-visits-daily.

33. Greenpeace, "Clicking Clean: Who Is Winning the Race to Build a Green Internet?" (Washington, DC: Greenpeace, 2017).

34. "AWS and Sustainability," Amazon.com, https://aws.amazon.com/about-aws/sustainability/#progress.

35. James Glanz, "Data Barns in a Farm Town, Gobbling Power and Flexing Muscle," *New York Times*, September 23, 2012, https://www.nytimes.com/2012/09/24/technology/data-centers-in-rural-washington-state-gobble-power.html.

36. Landau Associates, "Notice of Construction Application Supporting Information Report MWH-03/04/05/06 Data Center Quincy, Washington," June 6, 2018. https://ecology.wa.gov/DOE/files/74/7444566f-931f-4642-a864-5d3216494c09.pdf.

37. US Environmental Protection Agency, "Report Sections for the 2016 TRI National Analysis, January 2018, https://www.epa.gov/trinationalanalysis/report-sections-2016-tri-national-analysis.

38. Bay Area Air Quality Management District.

Taking It Offline: E-Commerce

1. Matt McFarland, "The case for almost never turning left while driving," *Washington Post*, April 9, 2014; Interview with Dan McMakin, United Parcel Service (UPS) spokesperson, February 14, 2018.

2. UPS spokesperson.

3. Antoine Gara, "Forbes Global 2000: The World's Largest Transportation Companies 2018," *Forbes*, June 6, 2018, https://www.forbes.com/sites/antoinegara/2018/06/06/forbes-global-2000-the-worlds-largest-transportation-companies.

4. IBISWorld, "Packing and Shipping Service Franchises in the US: US Industry Market Research Report" (Los Angeles: IBISWorld, 2017).

5. Statista, "Couriers and Local Delivery Service Providers' Global Market Share in 2017," https://www.statista.com/statistics/236309/market-share-of-global-express-industry.

6. United Parcel Service, "On the Leading Edge: UPS 2017 Corporate Sustainability Progress Report," (Atlanta, GA: 2017): https://sustainability.ups.com/media/2017_UPS_CSR.pdf.

7. Statista, "Number of Digital Shoppers in the United States from 2016 to 2021 (in Millions)," https://www.statista.com/statistics/183755/number-of-us-Internet-shoppers-since-2009.

8. Statista, "Retail E-Commerce Sales Worldwide from 2014 to 2021 (in Billion U.S. Dollars)," https://www.statista.com/statistics/379046/worldwide-retail-e-commerce-sales.

9. "Asia to account for more than half global eCommerce," *Inside Retail*, December 17, 2015, https://insideretail.hk/2015/12/17/asia-to-account-for-more-than-half-global-ecommerce/.

10. *Inside Retail.*

11. Sonja Kroll, "Retail E-Commerce Sales Rising; Evergage Expands into Europe," *Retail Tech News*, February 15, 2018. https://www.retailtechnews.com/2018/02/15/retail-e-commerce-sales-rising-evergage-expands-into-europe/

12. Rachel Kenyon, Fibre Box Association, email to author, February 6, 2018.

13. "Papermakers Vie for Southeast Asia Corrugated Cardboard Market," *Nikkei Asian Review*, April 18, 2017, https://asia.nikkei.com/Business/Business-Trends/Papermakers-vie-for-Southeast-Asia-corrugated-cardboard-market.

14. World Resources Institute, "Energy Efficiency in U.S. Manufacturing: The Case of Midwest Pulp and Paper Mills" (Washington, DC: World Resources Institute, 2013), 2.

15. US Energy Information Administration, "Food and Paper Account for 3.4% of Nation's Energy Use," July 5, 2013, https://www.eia.gov/todayinenergy/detail.php?id=11971.

16. Environmental Protection Administration, "Wastes—Resource Conservation—Common Waste and Materials—Paper Recycling—Frequent Questions," https://archive.epa.gov/wastes/conserve/materials/paper/web/html/faqs.html.

17. Rachel Kenyon.

18. American Chemistry Council, "2014 United States National Postconsumer Plastics Bottle Recycling Report," 2015.

19. World Economic Forum, "The New Plastics Economy: Rethinking the Future of Plastics," January 2016.

20. Elizabeth Weise, "Blue Bins Overflow with Amazon and Walmart Boxes. But

We're Actually Recycling Less," *USA Today*, June 8, 2018, https://www.usatoday.com/story/tech/news/2018/06/08/cardboard-recycling-rates-drop-shopping-amazon-walmart-surges/630967002.

21. Jamshid Laghaei, Ardeshir Faghri, and Mingxin Li, "Impacts of Home Shopping on Vehicle Operations and Greenhouse Gas Emissions: Multi-Year Regional Study," *International Journal of Sustainable Development and World Ecology* 23, no. 5 (2015).

22. David Schrank, Bill Eisele, Tim Lomax, and Jim Bak, "2015 Urban Mobility Scorecard," Texas A&M Transportation Institute and INRIX, August 2015, p. 1.

23. Orit Rotem Mindali and Jesse Weltevreden, "Transport Effects of E-Commerce: What Can Be Learned after Years of Research?" *Transportation* 40, no. 5 (2013): 5.

24. Ann Starodaj, "The Brewing Environmental Problem of Retail Returns," UPS Longitude, January 9, 2018, https://longitudes.ups.com/20877-2.

25. "Passenger Travel Facts and Figures 2016," US Department of Transportation, Bureau of Transportation Statistics, https://www.bts.gov/sites/bts.dot.gov/files/docs/PTFF%202016_2l.pdf.

26. Peter Dizikes, "The 6-percent solution: How corporations can reduce greenhouse-gas emissions through better planning," MIT News Office, November 8, 2010.

27. Andy Murdock, "The environmental cost of free two-day shipping," Vox, November 17, 2017.

28. José Holguín Veras et al., "The New York City Off-Hour Delivery Program: A Business and Community-Friendly Sustainability Program," *Interface* 48, no. 1 (2018): 1.

29. United Parcel Service and GreenBiz Research, "The Road to Sustainable Urban Logistics," 2017.

30. Florian Dost and Erik Maier, "E-Commerce Effects on Energy Consumption: A Multi-Year Ecosystem-Level Assessment," *Journal of Industrial Ecology* 22, no. 4 (2017).

31. Florian Dost, interview with author, February 8. 2018.

32. Kieren Mayers et al., "The Carbon Footprint of Games Distribution," *Journal of Industrial Ecology* 19, no. 3 (2014).

33. Kieren Mayers et al.

Silicon Valley: A Toxic Waste Dump? You Decide

1. Alexis C. Madrigal, "Not Even Silicon Valley Escapes History," *The Atlantic*, July 23, 2013.

2. EPA spokesperson, interview with author, December 21, 2017.

3. EPA, "Superfund Site: Fairchild Semiconductor Corp. (Mountain View Plant), Mountain View, CA," https://cumulis.epa.gov/supercpad/SiteProfiles/index.cfm?fuseaction=second.schedule&id=0901680.

4. Susanne Rust and Matt Drange, "Google Workers at Superfund Site Exposed," *SFGate*, March 19, 2013.

5. Zoë Schlanger, "Silicon Valley is home to more toxic Superfund sites than any-where else in the country," June 28, 2017, Quartz, https://qz.com/1017181/silicon-valley-pollution-there-are-more-superfund-sites-in-santa-clara-than-any-other-us-county/.

6. EPA spokesperson, interview with author, December 21, 2017.

7. Amanda Hawes, interview with author, February 13, 2018.

8. CDC, "Trichloroethylene Toxicity: What Are the Physiological Effects of Trichloroethylene?"; P. Tachachartvanich, R. Sangsuwan, H. S. Ruiz, S. S. Sanchez, K. A. Durkin, L. Zhang, M. T. Smith, "Assessment of the Endocrine-Disrupting Effects of Trichloroethylene and Its Metabolites Using in Vitro and in Silico Approaches," *Environmental Science and Technology* 52, no. 3 (2018): 1542–1550.

9. K. Claxton, M. Deane, G. Shaw, S. H. Swan, and M. Wrensch, "Pregnancy Outcomes in Santa Clara County 1980–1985," Epidemiological Studies and Surveillance Section, California Department of Health Services, 1988.

10. Stephen Stock, David Paredes, and Scott Pham, "Toxic Plumes: The Dark Side of Silicon Valley," NBC Bay Area, May 12, 2014.

11. EPA spokesperson.

12. Amanda Hawes.

13. Cam Simpson, "American Chipmakers Had a Toxic Problem. Then They Outsourced It," *Bloomberg Businessweek*, June 15, 2017.

14. Simpson.

Mining for Tech

1. Todd C. Frankel, "The Cobalt Pipeline: Tracing the Path from Deadly Hand-Dug Mines in Congo to Consumers' Phones and Laptops," *Washington Post*, September 30, 2016.

2. Frankel.

3. Apple, "Smelter and Refiner List," December 2017, https://www.apple.com/supplier-responsibility/pdf/Apple-Smelter-and-Refiner-List.pdf.

4. Todd C. Frankel, "Apple Cracks Down Further on Cobalt Supplier in Congo as Child Labor Persists," *Washington Post*, March 3, 2017.

5. Dr. Jana Hönke, Sarah Katz-Lavigne, "Cobalt isn't a conflict mineral," *Africa Is a Country*, September 11, 2018. https://africasacountry.com/2018/09/cobalt-isnt-a-conflict-mineral

6. Todd C. Frankel and Peter Whoriskey, "Tossed Aside in the 'White Gold' Rush: Indigenous People Are Left Poor as Tech World Takes Lithium from under Their Feet," *Washington Post*, December 19, 2016.

7. NASA Earth Observatory, Rare Snow in Atacama Desert, Chile, July 7, 2011. https://earthobservatory.nasa.gov/images/51312/rare-snow-in-atacama-desert-chile.

8. Frankel and Whoriskey.

9. Peter Whoriskey, "In Your Phone, In Their Air: A Trace of Graphite Is in Consumer Tech. In These Chinese Villages, It's Everywhere." *Washington Post*, October 2, 2016.

Vampire Power

1. Alan Meier, phone interview with author, April 1, 2016; Natural Resources Defense Council, "Home Idle Load: Device Wasting Huge Amounts of Electricity When Not in Active Use" (New York: Natural Resources Defense Council, 2015).
2. NRDC.
3. Energy Information Administration, "Power Sector Carbon Dioxide Emissions Fall below Transportation Sector Emissions" (Washington, DC: Energy Information Administration, 2017).
4. World Bank, "Electric Power Consumption (kWh per capita)," https://data .worldbank.org/indicator/EG.USE.ELEC.KH.PC.
5. World Bank, "Electric Power Consumption."
6. World Bank, "Electric Power Consumption."
7. Internet/Broadband Fact Sheet, Pew Research Center, February 5, 2018, http:// www.pewInternet.org/fact-sheet/Internet-broadband/.
8. NRDC, "Home Idle Load: Devices Wasting Huge Amounts of Electricity When Not in Active Use," https://www.nrdc.org/resources/home-idle-load-devices -wasting-huge-amounts-electricity-when-not-active-use.

The Tech We Throw Away

1. "Basic Information about Electronics Stewardship," Environmental Protection Agency, https://www.epa.gov/smm-electronics/basic-information-about -electronics-stewardship.
2. United Nations Environment Programme, *Waste Crime, Waste Risks: Gaps in Meeting the Global Waste Challenges*, (Nairobi, Kenya: United Nations Environment Programme, 2015): 7.
3. Devin N. Perkins, Marie-Noel Brune Drisse, Taipwa Nxele, and Peter D. Sly, "E-Waste: A Global Hazard," *Annals of Global Health* 80, no. 4 (2014): 286–295.
4. Perkins et al.
5. Perkins et al.
6. World Trade Organization, "China's Import Ban on Solid Waste Queried at Import Licensing Meeting," 2017, https://www.wto.org/english/news_e/news17 _e/impl_03oct17_e.htm.
7. United Nations University, "The Global E-Waste Monitor: Quantities, Flows and Resources" (Bonn, Germany: United Nations University, 2015), https://i.unu .edu/media/unu.edu/news/52624/UNU-1stGlobal-E-Waste-Monitor-2014-small .pdf.
8. Perkins et al.
9. Perkins et al.
10. Perkins et al.
11. Pure Earth, "Report from Agbogbloshie, Ghana," April 9, 2010, http://www .pureearth.org/blog/report-from-ghanas-agbogbloshie-e-wasteland.

12. Perkins et al.
13. Kristen Grant, Fiona C. Goldizen, Peter D. Sly, Marie-Noel Brune, Martin van den Berg, and Rosana E. Norman, "Health Consequences of Exposure to E-Waste: A Systematic Review," *Lancet Global Health* 1 (2013): 350–361.
14. Perkins et al.
15. Perkins et al.
16. Grant et al.
17. David Barboza, "Q. & A.: Adam Minter on Why China's Scrap Trade Helps the Environment," *New York Times*, November 29, 2013, https://sinosphere.blogs .nytimes.com/2013/11/29/q-a-adam-minter-on-why-chinas-scrap-trade-helps -the-environment.
18. United States Geological Survey, "Gold," https://minerals.usgs.gov/minerals/ pubs/commodity/gold/mcs-2017-gold.pdf.
19. Department of Energy, "Department of Energy, FY 2018 Congressional Budget Request," May 2017, https://www.energy.gov/sites/prod/files/2017/05/f34/ FY2018BudgetVolume3.pdf.
20. The Aluminum Association, "Aluminum Recycling," http://www.aluminum .org/sustainability/aluminum-recycling.
21. Apple, "Apple GiveBack: Turn the Device You Have into the One You Want," https://www.apple.com/shop/trade-in.
22. Best Buy, "Trade In or Recycle the Old Tech, Enjoy the New," December 26, 2017, https://corporate.bestbuy.com/trade-recycle-old-tech-enjoy-new.
23. Perkins et al.
24. Nedal Nassar, interview with author, November 21, 2017.
25. Nassar.
26. Basel Convention on the Control of Transboundary Movements of Hazardous Wastes and Their Disposal, March 22, 1989, http://www.basel.int/The Convention/Overview/tabid/1271/Default.aspx.
27. Bamako Convention on the Ban of the Import into Africa and the Control of Transboundary Movement and Management of Hazardous Wastes within Africa, January 30, 1991, https://au.int/en/treaties/bamako-convention-ban-import-africa -and-control-transboundary-movement-and-management.
28. European Union Network for the Implementation and Enforcement of Environmental Law, "Thousands of Tonnes of E-Waste Is Shipped Illegally to Nigeria Inside Used Vehicles," April 19, 2018, https://unu.edu/media-relations/releases/pip -press-release.html.
29. Nikkei Asian Review, "Japanese Companies Digging for Gold in Urban Waste," October 3, 2017, https://asia.nikkei.com/Business/Japanese-companies-digging -for-gold-in-urban-waste.
30. Nippon Telegraph and Telephone Corporation, "Tokyo 2020 Medal Project: Towards an Innovative Future for All," 2017, http://www.ntt.co.jp/news2017/1711efrz/ trqh171110a_13.html.

Food

1. Navin Ramankutty, Amato T. Evan, Chad Monfreda, and Jonathan A. Foley, "Farming the Planet: 1. Geographic Distribution of Global Agricultural Lands in the Year 2000," *Global Biogeochemical Cycles* 22, no. 1 (2008).
2. Jonathan Foley, "A Five Step Plan to Feed the World," *National Geographic*, 2014, https://www.nationalgeographic.com/foodfeatures/feeding-9-billion.
3. Climate Nexus, "Food, Agriculture and Climate Change," https://climatenexus .org/climate-issues/food/food-agriculture-and-climate-change.
4. Food and Agriculture Organization of the United Nations, "Tackling Climate Change through Livestock: A Global Assessment of Emissions and Mitigation Opportunities," 2013.
5. Julie Hecht, "Well That Stinks! Reporters Blow Cow Farts out of Proportion," *Scientific American*, February 13, 2014.
6. Food and Agriculture Organization of the United Nations, "How to Feed the World in 2050," 2009.

The Greediest Crop

1. Ross E. Alter, Hunter C. Douglas, Jonathan M. Winter, and Elfatih A. B. Eltahir, "Twentieth Century Regional Climate Change during the Summer in the Central United States Attributed to Agricultural Intensification," *Geophysical Research Letters* 45, no. 3 (February 16, 2018): 1586–1594.
2. Michael Pollan, *The Omnivore's Dilemma: A Natural History of Four Meals* (New York: Penguin Press, 2006), 23.
3. National Corn Growers' Association, "Corn Usage by Segment 2017," http://www.worldofcorn.com/#corn-usage-by-segment.
4. Jonathan Foley, "It's Time to Rethink America's Corn System," *Scientific American*, March 5, 2013.
5. United States Department of Agriculture, Census of Agriculture Historical Archive, http://usda.mannlib.cornell.edu/usda/AgCensusImages/1920/Individual _Crops.pdf; USDA, National Agriculture Statistics Service, https://quickstats.nass .usda.gov/results/22D644C3-2BB2-3E19-A2D3-2622679D6EE5; National Corn Growers' Association, "U.S. Corn Production," http://www.worldofcorn.com/ #us-corn-production-metric.
6. Pollan, 25–26.
7. Pollan, 25.
8. Cynthia Clampitt, *Midwest Maize; How Corn Shaped the U.S. Heartland* (Chicago: University of Illinois Press, 2015), 2.
9. Clampitt, 1.
10. Dave Merrill and Lauren Leatherby, "Here's How America Uses Its Land," *Bloomberg Businessweek*, July 31, 2018.
11. Christopher K. Wright and Michael C. Wimberly, "Recent Land Use Change in

the Western Corn Belt Threatens Grasslands and Wetlands," *Proceedings of the National Academy of Sciences* 110, no. 10 (2013): 4134–4139.

12. Wright and Wimberly.

13. Joseph Fargione, Jason Hill, David Tilman, Stephen Polasky, and Peter Hawthorne, "Land Clearing and the Biofuel Carbon Debt," *Science* 319, no. 29 (February 2008): 1235–1238.

14. FAO, "What Is Happening to Agrobiodiversity?" http://www.fao.org/docrep/007/y5609e/y5609e02.htm.

15. Pollan, 45.

16. Wayne Hoffman, Jan Beyea, and James Cook, "Ecology of Agricultural Monocultures: Some Consequences for Biodiversity in Biomass Energy Farms," in *Proceedings of the 2nd Biomass Conference of the Americas* (Golden, CO: National Renewable Energy Laboratory, 1995), 1618–1627.

17. Michael Pollan, *The Botany of Desire: A Plant's-Eye View of the World* (New York: Random House, 2001), 226.

18. Miguel A. Altieri, "The Ecological Role of Biodiversity in Agroecosystems," in *Invertebrate Biodiversity as Bioindicators of Sustainable Landscapes: Practical Use of Invertebrates to Assess Sustainable Land Use*, ed. Maurizio G. Paoletti (Amsterdam: Elsevier, 1999), 19–31.

19. Sean L. Maxwell, Richard A. Fuller, Thomas M. Brooks, and James E. M. Watson, "Biodiversity: The Ravages of Guns, Nets, and Bulldozers," *Nature* 536, no. 7615 (August 10, 2016): 143–145.

20. Pollan, *The Omnivore's Dilemma*, 149.

21. Mark Kinver, "Crop Diversity Decline 'Threatens Food Security'," BBC News, March 3, 2014.

22. Mark Bittman, "A Sustainable Solution for the Corn Belt," *New York Times*, November 18, 2014; Craig Cox, Andrew Hug, and Nils Bruzelius, "Losing Ground," Environmental Working Group, April 2011.

23. USDA, Natural Resources Conservation Service, "Rotations for Soil Fertility: Small Scale Solutions for Your Farm," 2009.

24. David Tilman, "The Greening of the Green Revolution," *Nature* 396 (November 19, 1998): 211–212, at 211.

25. Tim Harford, "How Fertiliser Helped Feed the World," BBC News, January 2, 2017.

26. Sam Wood and Annette Cowie, "A Review of Greenhouse Gas Emission Factors for Fertiliser Production," IEA Bioenergy Task, 2004.

27. Foley.

28. NOAA, "Gulf of Mexico 'Dead Zone' Is the Largest Ever Measured," press release, August 2, 2017, http://www.noaa.gov/media-release/gulf-of-mexico-dead-zone-is-largest-ever-measured.

29. Environmental Protection Agency, "Mississippi River/Gulf of Mexico Watershed Nutrient Task Force, 2015 Report to Congress," https://www.epa.gov/sites/production/files/2015-10/documents/htf_report_to_congress_final_-_10.1.15.pdf.

30. Donald Scavia, Isabella Bertani, Daniel R. Obenour, R. Eugene Turner, David R. Forrest, and Alexey Katin, "Ensemble Modeling Informs Hypoxia Management in the Northern Gulf of Mexico," *Proceedings of the National Academy of Sciences* 114, no. 33 (August 15, 2017): 8823–8828.

31. E. Sinha, A. M. Michalak, and V. Balaji, "Eutrophication Will Increase during the 21st Century as a Result of Precipitation Changes," *Science* 357, no. 6349 (July 28, 2017): 405–408.

32. K. J. Van Meter, P. Van Cappellen, and N. B. Basu, "Legacy Nitrogen May Prevent Achievement of Water Quality Goals in the Gulf of Mexico," *Science* 360, no. 3587 (March 22, 2018): 427–430.

33. M. H. Ward, B. A. Kilfoy, P. J. Weyer, K. E. Anderson, A. R. Folsom, and J. R. Cerhan, "Nitrate Intake and the Risk of Thyroid Cancer and Thyroid Disease," *Epidemiology* 21, no. 3 (May 2010): 389–395.

34. Peter J. Weyer, James R. Cerhan, Burton C. Kross, George R. Hallberg, Jiji Kantamneni, George Breuer, Michael P. Jones, Wei Zheng, and Charles F. Lynch, "Municipal Drinking Water Nitrate Level and Cancer Risk in Older Women: The Iowa Women's Health Study," *Epidemiology* 12, no. 3 (2001): 327–338.

35. Rena R. Jones, Peter J. Weyer, Curt T. DellaValle, Maki Inoue-Choi, Kristin E. Anderson, Kenneth P. Cantor, Stuart Krasner, Kim Robien, Laura E. Beane Freeman, Debra T. Silverman, and Mary H. Ward, "Nitrate from Drinking Water and Diet and Bladder Cancer among Postmenopausal Women in Iowa," *Environmental Health Perspectives* 124, no. 11 (November 2016): 1751–1758.

36. Environmental Working Group, Tap Water Database, https://www.ewg.org/tapwater.

37. Foley.

Wasting Away

1. USDA Economic Research Service, "The Estimated Amount, Value, and Calories of Postharvest Food Losses at the Retail and Consumer Levels in the United States," Economic Information Bulletin Number 121, February 2014.

2. Dana Gunders, "Wasted: How America Is Losing up to 40 Percent of Its Food from Farm to Fork to Landfill," NDRC, August 16, 2017.

3. Gunders.

4. WRAP, "Dr Liz Goodwin to Step Down as WRAP CEO," February 8, 2016, http://www.wrap.org.uk/content/dr-liz-goodwin-step-down-wrap-ceo.

5. Liz Goodwin, interview with author, May 22, 2018.

6. Office of Technological Assessment, "Open Shelf-Life Dating of Food," August 1979.

7. Office of Technological Assessment.

8. Office of Technological Assessment.

9. NRDC, "The Dating Game: How Confusing Labels Land Billions of Pounds of Food in the Trash," September 2013.

10. Tony Benjamin, "Milk labeling aids shopper," *Boca Raton News*, October 10, 1972, https://news.google.com/newspapers?nid=1291&dat=19721010&id=wMJTAAA AIBAJ&sjid=No0DAAAAIBAJ&pg=4279,4059788.

11. NRDC.

12. NRDC.

13. Food Marketing Institute, "U.S. Grocery Shopper Trends 2015" (2015), https://www.fmi.org/our-research/research-reports/u-s-grocery-shopper-trends.

14. ReFED, "Standardized Date Labeling," https://www.refed.com/solutions/standardized-date-labeling.

15. Jean C. Buzby, Hodan F. Wells, and Jeffrey Hyman, "The Estimated Amount, Value, and Calories of Postharvest Food Losses at the Retail and Consumer Levels in the United States," Economic Research Service, USDA, Economic Information Bulletin Number 121, February 2014.

16. Buzby, Wells, and Hyman.

17. Gunders.

18. JoAnne Berkenkamp, Darby Hoover, and Yerina Mugica, "Food Matters: What Food We Waste and How We Can Expand the Amount of Food We Rescue," NRDC, October 24, 2017.

19. Gunders.

20. ReFED, "A Roadmap to Reduce US Food Waste by 20 Percent," (2016).

21. Environmental Protection Agency, "America's Food Waste Problem," April 22, 2016.

22. US General Accounting Office, "Food Waste: An Opportunity to Improve Resource Use," September 16, 1977.

23. Gunders.

24. ReFED, A Roadmap to Reduce US Food Waste by 20 Percent; Gunders.

25. Gunders, 14.

26. Gunders.

27. USDA National Agricultural Statistics Service, "Vegetables 2014 Summary."

28. USDA National Agricultural Statistics Service.

29. Gunders.

30. Gunders.

31. Reilly Brock, interview with author, May 14, 2018.

32. Imperfect Produce, https://www.imperfectproduce.com.

33. Emily Atkin, "Does Your Box of 'Ugly' Produce Really Help the Planet? Or Hurt It?" *The New Republic*, January 11, 2019, https://newrepublic.com/article/152596/hungry-harvest-box-ugly-produce-help-planet-or-hurt-it.

34. Jenny Gustavsson et al., "Global Food Losses and Food Waste," UN FAO (2011) www.fao.org/food-loss-and-food-waste/en.

35. USDA Economic Research Service.

36. USDA Economic Research Service.

37. EPA, "United States 2030 Food Loss and Waste Reduction Goal," 2015.

38. Gunders.

Organic Food: How Good Is It?

1. USDA Agricultural Marketing Service, "Organic Labeling Standards," https://www.ams.usda.gov/grades-standards/organic-labeling-standards.
2. Christie Wilcox, "Mythbusting 101: Organic Farming > Conventional Agriculture," *Scientific American*, July 18, 2011.
3. USDA Agricultural Marketing Service, Science and Technology Program, "Pesticide Data Program, Annual Summary, Calendar Year 2016," February 2018, pp. ix–x.
4. Crystal Smith-Spangler, Margaret L. Brandeau, Grace E. Hunter, et al., "Are Organic Foods Safer or Healthier than Conventional Alternatives? A Systematic Review," *Annals of Internal Medicine* 157, no. 5 (September 4, 2012): 348–366.
5. C. Bahlai, Y. Xue, C. McCreary, A. Schaafsma, and R. Hallett, "Choosing Organic Pesticides over Synthetic Pesticides May Not Effectively Mitigate Environmental Risk in Soybeans," *PLOS ONE* 5, no. 6 (June 22, 2010).
6. Visit to Rogers Orchards, Southington, Connecticut, June 26, 2018.
7. Michael Biltonen, interview with author during Rogers Orchards visit.
8. Washington Apple Commission, https://bestapples.com.
9. Michael Pollan, *The Omnivore's Dilemma: A Natural History of Four Meals* (New York: Penguin Press, 2006).
10. H. L. Tuomisto, D. Hodge. P. Riordana, and D. W. Macdonald, "Does Organic Farming Reduce Environmental Impacts? A Meta-Analysis of European Research," *Journal of Environmental Management* 112 (December 15, 2012): 309–320.
11. Pietro Barbieri, Sylvain Pellerin, and Thomas Nesme, "Comparing Crop Rotations between Organic and Conventional Farming," *Scientific Reports* 7 (2017): 13761.
12. Verena Seufert and Navin Ramankutty, "Many Shades of Gray: The Context-Dependent Performance of Organic Agriculture," *Science Advances* (March 10, 2017).
13. Tuomisto et al.
14. Seufert and Ramankutty.
15. Seufert and Ramankutty.
16. H. Willer and J. Lernoud, eds., "The World of Organic Agriculture: Statistics and Emerging Trends 2017" (Frick, Switzerland: Research Institute of Organic Agriculture FiBL, IFOAM–Organics International, 2017), 182.
17. Karl-Heinz Erb, Christian Lauk, Thomas Kastner, Andreas Mayer, Michaela C. Theurl, and Helmut Haberl, "Exploring the Biophysical Option Space for Feeding the World without Deforestation," *Nature Communications* 7 (April 19, 2016): 11382.
18. Mark Bittman, "A Sustainable Solution for the Corn Belt," *New York Times*, November 18, 2014.
19. Tuomisto et al.
20. Adam S. Davis. Jason D. Hill, Craig A. Chase, et al., "Increasing Cropping System Diversity Balances Productivity, Profitability, and Environmental Health," *PLOS ONE* 7, no. 10 (October 2012).

21. World Health Organization, "World Hunger Again on the Rise, Driven by Conflict and Climate Change, New UN Report Says," press release, September 15, 2017.

How Far Our Food Goes

1. FDA, "Advancing the Safety of Imported Food," 2017; David Karp, "Most of America's Fruit Is Now Imported. Is That a Bad Thing?" *New York Times*, March 13, 2018.
2. Karp.
3. Mary Beth Albright, "The Five Best Fruits You're Not Eating," *National Geographic*, July 21, 2014.
4. Karp.
5. Karp.
6. A. Paxton, "The Food Miles Report: The Dangers of Long Distance Food Transport" (London: Safe Alliance, 1994).
7. Christopher L. Weber and H. Scott Matthews, "Food Miles and the Relative Climate Impacts of Food Choices in the U.S.," *Environmental Science and Technology* 42, no. 10 (2008): 3508–3513.
8. Misak Avetisyan, Thomas Hertel, and Gregory Sampson, "Is Local Food More Environmentally Friendly? The GHG Emissions Impact of Consuming Imported versus Domestically Produced Food," *Environmental and Resource Economics* 58, no. 3 (July 2014).
9. Sarah Murray, *Moveable Feasts: From Ancient Rome to the 21st Century, the Incredible Journeys of the Food We Eat* (New York: St. Martin's Press, 2007), 402.
10. "Food Product Environmental Footprint Literature Summary: Food Transportation," Center for Sustainable Systems, University of Michigan report for State of Oregon, Department of Environmental Quality, 2017.
11. "Good Food?" *The Economist*, December 7, 2006.
12. Sam Bloch, "2017's Natural Disasters Cost American Agriculture over $5 billion," *New Food Economy*, January 4, 2018.
13. Murray, p. 402; "Food Product Environmental Footprint Literature Summary: Food Transportation"; Avetisyan, Hertel, and Sampson; Adrian Williams, "How the Myth of Food Miles Hurts the Planet," *The Observer*, March 23, 2008.
14. Stefanie Böge, "The Well-Travelled Yogurt Pot: Lessons for New Freight Transport Policies and Regional Production," *World Transport Policy and Practice* 1, no. 1 (1995): 7–11.

A Sea of Troubles

1. Food and Agricultural Organization of the United Nations, "The State of World Fisheries and Aquaculture," 2018.
2. David J. Agnew, John Pearce, Ganapathiraju Pramod, Tom Peatman, Reg Watson, John R. Beddington, and Tony J. Pitcher, "Estimating the Worldwide Extent of Illegal Fishing," *PLOS ONE* 4, no. 2 (February 2009).

3. Paul Greenberg, *American Catch: The Fight for Our Local Seafood* (New York: Penguin Books, 2015).

4. Natural Resources Defense Council, "Net Loss: The Killing of Marine Mammal in Foreign Fisheries," January 2014.

5. NOAA Fisheries.

6. Brad Plumer, "How the US Stopped Its Fisheries from Collapsing," Vox, May 8, 2014.

7. NOAA Fisheries, "2017 Report to Congress on the Status of U.S. Fisheries," 2017.

8. David A. Kroodsma, Juan Mayorga, Timothy Hochberg, Nathan A. Miller, Kristina Boerder, Francesco Ferretti, Alex Wilson, Bjorn Bergman, Timothy D. White, Barbara A. Block, Paul Woods, Brian Sullivan, Christopher Costello, and Boris Worm, "Tracking the Global Footprint of Fisheries," *Science* 359, no. 6378 (February 23, 2018): 904–908.

9. Food and Agricultural Organization of the United Nations.

10. Dan Laffoley and John M. Baxter, eds., "Explaining Ocean Warming: Causes, Scale, Effects, and Consequences," International Union for the Conservation of Nature, September 2016.

11. J. M. Sunday, A. E. Bates, and N. K. Dulvy, "Thermal Tolerance and the Global Redistribution of Animals," *Nature Climate Change* 2 (2012): 686–690; E. S. Poloczanska, C. J. Brown, W. J. Sydeman, et al., "Global Imprint of Climate Change on Marine Life," *Nature Climate Change* 3 (2013): 919–925.

12. Rebecca G. Asch, "Climate Change and Decadal Shifts in the Phenology of Larval Fishes in the California Current Ecosystem," *Proceedings of the National Academy of Sciences* 112, no. 30 (2015): E4065–E4074.

13. James W. Morley, Rebecca L. Selden, Robert J. Latour, Thomas L. Frölicher, Richard J. Seagraves, and Malin L. Pinsky, "Projecting Shifts in Thermal Habitat for 686 Species on the North American Continental Shelf," *PLOS ONE* 13, no. 5 (2018).

14. Terry P. Hughes, Kristen D. Anderson, Sean R. Connolly, et al., "Spatial and Temporal Patterns of Mass Bleaching of Corals in the Anthropocene," *Science* 359, no. 6371 (January 5, 2018): 80–83.

15. "Coral Reefs: The Ocean's Larder," BBC News, February 20, 2013.

16. Captain John Smith, "A Description of New-England" (1616), in *The Complete Works of Captain John Smith (1580–1621)*, 3 vols., ed. Philip L. Barbour (Chapel Hill, NC: University of North Carolina Press, 1986), 1:347.

17. "Will the Fish Return?" American Museum of Natural History, https://www.amnh.org/exhibitions/permanent-exhibitions/irma-and-paul-milstein-family-hall-of-ocean-life/resources-for-educators/conservation/will-the-fish-return.

18. Marianne Lavelle, "Collapse of New England's Iconic Cod Tied to Climate Change," *Science*, October 29, 2015.

19. Patrick Whittle, "Why New England's Cod Catch Is at an All-Time Low," *Boston Globe*, January 14, 2018.

20. Whittle.

21. Tatiana Schlossberg, "Acidic Ocean Leads to Warped Skeletons for Young Coral," *New York Times*, February 19, 2016.

22. Ulf Riebesell, Nicole Aberle-Malzahn, Eric P. Achterberg, et al., "Toxic Algal Bloom Induced by Ocean Acidification Disrupts the Pelagic Food Web," *Nature Climate Change* 8 (2018): 1082-1086; National Oceanic and Atmospheric Administration, "Why Do Harmful Algal Blooms Occur?" https://oceanservice.noaa.gov/facts/why_habs.html.

23. Denise Breitburg, Lisa A. Levin, Andreas Oschlies, et al., "Declining Oxygen in the Global Ocean and Coastal Waters," *Science* 359, no. 6371 (January 5, 2018).

24. Raquel Vaquer-Sunyer and Carlos M. Duarte, "Thresholds of Hypoxia for Marine Biodiversity," *Proceedings of the National Academy of Sciences* 105, no. 40 (October 7, 2008): 15452–15457.

25. Bethan C. O'Leary, Marit Winther-Janson, John M. Bainbridge, et al., "Effective Coverage Targets for Ocean Protection," *Conservation Letters* 9, no. 6 (March 21, 2016): 398–404.

26. Nicola Jones, "A Growing Call for International Marine Reserves," Yale Environment 360, September 29, 2016.

27. Emiko Terazono, "Farmed Fish on Course to Overtake Wild Catch in 2019," *Financial Times*, May 28, 2017.

28. Ray Hilborn, Jeannette Banobi, Stephen J. Hall, Teresa Pucylowski, and Timothy E. Walsworth, "The Environmental Cost of Animal Source Foods," *Frontiers in Ecology and the Environment* 16, no. 6 (August 2018): 329–335.

29. E. Pikitch, P. D. Boersma, I. L. Boyd, et al., "Little Fish, Big Impact: Managing a Crucial Link in Ocean Food Webs" (Washington, DC: Lenfest Ocean Program, April 2012).

30. Theingi Aung, Jim Halsey, Daan Kromhout, et al., "Associations of Omega-3 Fatty Acid Supplement Use with Cardiovascular Disease Risks: Meta-Analysis of 10 Trials Involving 77 917 Individuals," *JAMA Cardiology* 3, no. 3 (2018): 225–233.

31. Henry Fountain, "Too Many Small Fish Are Caught, Report Says," *New York Times*, April 2, 2012.

32. Halley E. Froehlich, Nis Sand Jacobsen, Timothy E. Essington, et al., "Avoiding the Ecological Limits of Forage Fish for Fed Aquaculture," *Nature Sustainability* 1 (2018): 298–303.

33. Raphaëla Le Gouvello and François Simard, "Durabilité des aliments pour le poisson en aquaculture: Réflexions et recommandations sur les aspects technologiques, économiques, sociaux et environnementaux," International Union of Conservation Scientists, 2017.

34. Andrew Jacobs, "China's Appetite Pushes Fisheries to the Brink," *New York Times*, April 30, 2017.

35. Jim Wickens, "How Vital Fish Stocks in Africa Are Being Stolen from Human Mouths to Feed Pigs and Chickens on Western Factory Farms," *The Independent*, September 17, 2016.

36. Jacobs.
37. Greenberg, p. 2.
38. Terry Gross, "Interview with Paul Greenberg," *Fresh Air*, NPR, July 1, 2014.
39. Douglas J. McCauley, Malin L. Pinsky, Stephen R. Palumbi, et al., "Marine Defaunation: Animal Loss in the Global Ocean," *Science* 327, no. 6219 (January 16, 2015).
40. Becky Ferreira, "The Marine Biologist Using Big Data to Protect Ocean Wildlife," Vice, December 8, 2017.
41. https://labs.eemb.ucsb.edu/mccauley/doug/publications/genome_Douglas _McCauley_Full_20111229202245.txt; https://www.fitbit.com/user/39K34D.
42. Douglas McCauley, interview with author, June 22, 2018.
43. Tatiana Schlossberg, "A Plan to Give Whales and Other Ocean Life Some Peace and Quiet," *New York Times*, June 3, 2016.
44. Schlossberg.
45. Schlossberg.

Fashion

1. Linda Greer, interview with author, September 14, 2018.
2. James Conca, "Making Climate Change Fashionable: The Garment Industry Takes On Global Warming," *Forbes*, December 3, 2015.
3. Kimberly Cutter, "On Thin Ice: Can the Fashion Industry Help Save the Planet?" *Marie Claire*, August 18, 2016.
4. Alden Wicker, "We Have No Idea How Bad Fashion Actually Is for the Environment," *Racked*, March 15, 2017.
5. Kendra Pierre-Louis and Hiroko Tabuchi, "Want Cleaner Air? Try Using Less Deodorant," *New York Times*, February 16, 2018. https://www.nytimes.com/2018/02/16/climate/perfume-pollution-smog.html.

Thirsty for Denim

1. John Mack Faragher and Robert V. Hine, *A People's History of the American West: A New Interpretive History* (New Haven, CT: Yale University Press, 2000), 249.
2. Natural Resources Defense Council, "Clearing Up Your Choices on Cotton," 2011.
3. Paulina Szmydke-Cacciapalle, *Making Jeans Green: Linking Sustainability, Business and Fashion* (Routledge, 2018), 53 (Kindle edition).
4. Leslie A. Meyer, "Cotton and Wool Outlook, USDA Economic Research Service, March 12, 2019. https://www.ers.usda.gov/webdocs/publications/92565/cws-19c.pdf?v=3160.8.
5. Natural Resources Defense Council, "Clearing Up Your Choices on Cotton."
6. Richard Blackburn, ed., *Sustainable Textiles: Life Cycle and Environmental Impact* (Cambridge: Woodhead, 2009), 34; Szmydke-Cacciapalle, p. 40.

7. Szmydke-Cacciapalle, p. 38

8. Natural Resources Defense Council, "Clearing Up Your Choices on Cotton."

9. "Principles and Criteria," Better Cotton Initiative, https://bettercotton.org/about-better-cotton/better-cotton-standard-system/production-principles-and-criteria/.

10. Szmydke-Cacciapalle, p. 36.

11. World Wildlife Fund, "Living Waters: Conserving the Source of Life," 2002, p. 10.

12. "Freshwater Crisis," National Geographic, https://www.nationalgeographic.com/environment/freshwater/freshwater-crisis/

13. Szmydke-Cacciapalle, p. 37.

14. Elisa Schaar, "Central Asia's Dead Sea," Harvard International Review, September 6, 2001.

15. Richard Pomfret, "State-Directed Diffusion of Technology: The Mechanization of Cotton Harvesting in Soviet Central Asia," Journal of Economic History 62, no. 1 (2002): 170–188.

16. "Cotton Production at Aral Sea, Uzbekistan, and Turkmenistan," Environmental Justice Atlas, https://ejatlas.org/conflict/the-aral-sea-dried-due-to.

17. Karen Bennett, "Disappearance of the Aral Sea," World Resources Institute, May 23, 2008.

18. Bennett.

19. Dene-Hern Chen, "The Country That Brought a Sea Back to Life," BBC, July 23, 2018.

20. Rustam Qobil, "Waiting for the Sea," BBC, https://www.bbc.com/news/resources/idt-a0c4856e-1019-4937-96fd-8714d70a48f7.

21. Szmydke-Cacciapalle, p. 36.

22. Deb Berlin, "My Jeans Are Very Thirsty!" EPA Blog, June 14, 2010, https://blog.epa.gov/2010/06/14/my-jeans-are-very-thirsty.

23. Szmydke-Cacciapalle, p. 73.

24. Szmydke-Cacciapalle, pp. 75–76.

25. Subramanian Senthilkannan Muthu, ed., Sustainability in Denim (Oxford: Woodbridge Publishing, 2017), 39–41.

26. Greenpeace, "The Dirty Secret behind Jeans and Bras," December 1, 2010.

27. UN Environment Programme, "Options for Decoupling Economic Growth from Water Use and Water Pollution," May 2016, p. 2.

28. Alexandra S. Richey et al., "Quantifying Renewable Groundwater Stress with GRACE," AGU Water Resources Research 51, no. 7 (2015): 5217–5238.

29. Szmydke-Cacciapalle, p. 64.

30. Levi's, "Give More. Take Less," https://www.levi.com/US/en_US/features/sustainability.

31. Levi's, "How We Make Jeans with Less Water," https://www.levi.com/US/en_US/blog/article/how-we-make-jeans-with-less-water/.

32. Levi's, "Give More. Take Less."

33. Nic McCormack, "What Goes Into Making an Earth-Friendly $68 Pair of Jeans," Bloomberg, October 6, 2017, https://www.bloomberg.com/news/articles/ 2017-10-06/what-goes-into-making-everlane-s-eco-friendly-jeans.
34. Everlane spokespeople, email with author, October 29, 2018.
35. Lauren Sommer and Jason Margolis, "The Water in Your Jeans: How Two Consumer Products Giants Are Cutting Back on Water Use," KQED, January 12, 2018.
36. Levi's, "Give More. Take Less."

Athleisure Forever!

1. Food and Agriculture Organization of the United Nations, International Cotton Advisory Committee, "A Summary of the World Apparel Fiber Consumption Survey, 2005–2008," 2011, p. 3.
2. Deborah Drew and Genevieve Yehounme, "The Apparel Industry's Environmental Impact in 6 Graphics," World Resource Institute, July 5, 2017.
3. Paulina Szmydke-Cacciapalle, *Making Jeans Green: Linking Sustainability, Business and Fashion* (Routledge, 2018), 53 (Kindle edition), 54.
4. Austin K. Baldwin, "Occurrence and Potential Risk of Microplastics in Lake Mead and the Delaware River," United States Geological Survey, https://www .usgs.gov/centers/id-water/science/occurrence-and-potential-risk-microplastics -lake-mead-and-delaware-river?qt-science_center_objects=0#qt-science _center_objects.
5. Mark A. Browne et al., "Accumulation of Microplastic on Shorelines Worldwide: Sources and Sinks," *Environmental Science and Technology* 45, no. 21 (2011): 9175–9179.
6. Courtney Humphries, "Freshwater's Macro Microplastic Problem," PBS *Nova*, May 11, 2017.
7. USGS, NPS, "Microplastics Are Everywhere!" Great Lakes Network Resource Brief, April 4, 2017.
8. Imogen E. Napper, Richard C. Thompson, "Release of synthetic microplastic plastic fibres from domestic washing machines: Effects of fabric type and washing conditions," *Marine Pollution Bulletin*, Vol.112, Iss. 1–2, November 15, 2016, pp. 39–45.
9. N. L. Hartline, N. J. Bruce, S. N. Karba, et al., "Microfiber Masses Recovered from Conventional Machine Washing of New or Aged Garments," *Environmental Science and Technology* 50, no. 21 (2016): 11532–11538.
10. Hartline et al.
11. Humphries.
12. Austin K. Baldwin, Steven R. Corsi, and Sherri A. Mason, "Plastic Debris in 29 Great Lakes Tributaries: Relations to Watershed Attributes and Hydrology," *Environmental Science and Technology* 50, no. 19 (2016): 10377–10385.

13. Julien Boucher and Damien Friot, "Primary Microplastics in the Ocean: A Global Evaluation of Sources," International Union for Conservation of Nature, 2017, p. 21.
14. Rachid Dris, Johnny Gasperi, Vincent Rocher, et al., "Microplastic Contamination in an Urban Area: Case of Greater Paris," SETAC (Society of Environmental Toxicology and Chemistry) Europe 2015, May 2015, Barcelona, Spain.
15. Andrea Thompson, "Earth Has a Hidden Plastic Problem—Scientists Are Hunting It Down," Scientific American, August 13, 2018.
16. Patagonia, "Recycled Polyester," https://www.patagonia.com/recycled-polyester.html.
17. Patagonia, "An Update on Microfiber Pollution," February 3, 2017, https://www.patagonia.com/blog/2017/02/an-update-on-microfiber-pollution; Patagonia, "What You Can Do about Microfiber Pollution," June 26, 2017, https://www.patagonia.com/blog/2017/06/what-you-can-do-about-microfiber-pollution.
18. Annie Gullingsrud, Fashion Fibers: Designing for Sustainability (New York: Fairchild Books, 2017), 255.
19. Emily Chung, "Arctic Sea Ice Jammed with Plastics from Pacific Garbage Patch," CBC, April 24, 2018; Austin Baldwin; Austin Ramzy, "A Remote Pacific Island Awash in Tons of Trash," New York Times, May 16, 2017; Sarah Gibbens, "Plastic Bag Found at the Bottom of World's Deepest Ocean Trench," National Geographic, May 11, 2018; Matthew Taylor, "Antarctica: Plastic Contamination Reaches Earth's Last Wilderness," The Guardian, June 6, 2018.

Fast Fashion, but Going Nowhere

1. H&M, "History," https://about.hm.com/en/about-us/history.html.
2. "The World's 25 Largest Apparel Companies in 2018," Forbes, June 6, 2018.
3. Elizabeth Paton, "H&M, a Fashion Giant, Has a Problem: $4.3 Billion in Unsold Clothes," New York Times, March 27, 2018.
4. National Resources Defense Council, "The Textile Industry Leaps Forward with Clean by Design: Less Environmental Impact with Bigger Profits," 2015.
5. Nathalie Remy, Eveline Speelman, and Steven Swartz, "Style That's Sustainable: A New Fast-Fashion Formula," McKinsey & Co., October 2016.
6. Remy, Speelman, and Swartz.
7. Samantha Putt Del Pino, Eliot Metzger, Deborah Drew, and Kevin Moss, "The Elephant in the Boardroom: Why Unchecked Consumption Is Not an Option in Tomorrow's Markets," World Resources Institute, March 2017.
8. Remy, Speelman, and Swartz.
9. Remy, Speelman, and Swartz.
10. Environmental Protection Agency, Facts and Figures about Materials, Waste and Recycling.
11. EPA.
12. Remy, Speelman, and Swartz.

13. Waste and Resources Action Programme, "Valuing Our Clothes: The Cost of UK Fashion," July 2017.
14. Lucy Siegle, "Am I a Fool to Expect More than Corporate Greenwashing," *The Guardian*, April 2, 2016.
15. Paulina Szmydke-Cacciapalle, *Making Jeans Green: Linking Sustainability, Business and Fashion* (Routledge, 2018), 53 (Kindle edition), 99.
16. Andrew Brooks, interview with author, May 17, 2017.
17. "Measuring Fashion: Global Impact Study," Quantis/Climate Works.

It's Not Wood, It's Rayon

1. "Rayon," Encyclopædia Britannica.
2. Textile Exchange, "Preferred Fiber Market Report," 2016.
3. Textile World, "Man-Made Fibers Continue to Grow," February 3, 2015.
4. Agency for Toxic Substances and Disease Registry, "Carbon Disulfide Fact Sheet."
5. Rob Schmitz, "China Shuts Down Tens of Thousands of Factories in Unprecedented Pollution Crackdown," NPR, October 23, 2017.
6. Nithin Coca, "Despite Government Pledges, Ravaging of Indonesia's Forests Continues," *Yale Environment 360*, March 22, 2018.
7. SCS Global Services, "Executive Summary: Life Cycle Assessment Comparing Ten Sources of Manmade Cellulose Fiber," October 6, 2017.
8. D. G. McCullough, "Deforestation for Fashion: Getting Unsustainable Fabrics out of the Closet," *The Guardian*, April 25, 2014.
9. Rainforest Action Network, "Out of Fashion," https://www.ran.org/issue/out_of_fashion.
10. SCS Global Services, p. 12.
11. SCS Global Services, p. 9.
12. Federal Trade Commission, "FTC Charges Companies with 'Bamboo-zling' Consumers with False Product Claims," August 11, 2009, https://www.ftc.gov/news-events/press-releases/2009/08/ftc-charges-companies-bamboo-zling-consumers-false-product-claims.
13. Federal Trade Commission, "How to Avoid Bamboozling Your Customers," August 2009, https://www.ftc.gov/tips-advice/business-center/guidance/how-avoid-bamboozling-your-customers.
14. Federal Trade Commission, "FTC Warns 78 Retailers, Including Wal-Mart, Target, and Kmart, to Stop Labeling and Advertising Rayon Textile Products as 'Bamboo,'" February 3, 2010, https://www.ftc.gov/news-events/press-releases/2010/02/ftc-warns-78-retailers-including-wal-mart-target-kmart-stop.
15. Federal Trade Commission, "Four National Retailers Agree to Pay Penalties Totaling $1.26 Million for Allegedly Falsely Labeling Textiles as Made of Bamboo, While They Actually Were Rayon," January 3, 2013, https://www.ftc.gov/news-events/press-releases/2013/01/four-national-retailers-agree-pay-penalties-totaling-126-million.

16. National Resources Defense Council, "Not All Bamboo Is Created Equal," 2011.
17. Changing Markets Foundation, "Dirty Fashion: How Pollution in the Global Textiles Supply Chain Is Making Viscose Toxic," 2017, p. 18.
18. SCS Global Services, p. 10.
19. Changing Markets Foundation, p. 18.
20. Changing Markets Foundation, "Dirty Fashion: On track for transformation," 2018. http://changingmarkets.org/wp-content/uploads/2018/07/Dirty_Fashion_on _track_for_transformation.pdf.
21. Changing Markets Foundation, "Dirty Fashion Revisited: Spotlight on a Polluting Viscose Giant," 2018, pp. 6, 25, 7.

The Yarn That Makes a Desert

1. "Gobi," Encyclopædia Britannica online.
2. Timothy M. Kusky, Geological Hazards: A Sourcebook (Westport, CT: Greenwood Press, 2003), 183; Josh Haner, Edward Wong, Derek Watkins, and Jeremy White, "Living in China's Expanding Deserts," New York Times, October 24, 2016.
3. E. Papot, "Fashion Accessory: The Shawl," Napoleon.org, https://www.napoleon .org/en/magazine/napoleonic-pleasures/fashion-accessory-the-shawl.
4. Papot.
5. Pearly Jacob, "Mongolia: Herders Caught between Cashmere and Climate Change," Eurasianet, June 6, 2012.
6. Evan Osnos, "Your Cheap Sweater's Real Cost," Chicago Tribune, December 16, 2006.
7. Rebecca Mead, "The Crisis in Cashmere," New Yorker, February 1, 1999.
8. Jacob.
9. Rob Schmitz, "How Your Cashmere Sweater Is Decimating Mongolia's Grasslands," NPR, December 9, 2016.
10. Schmitz.
11. Lucy Siegle, "Should I Worry about Cheap Cashmere?" The Guardian, December 7, 2014.
12. US AID, "A Value Chain Analysis of the Mongolian Cashmere Industry," May 2005.
13. Lisa Friedman, Kendra Pierre-Louis, and Somini Sengupta, "The Meat Question, by the Numbers," New York Times, January 25, 2018.
14. Statista, "Number of Cattle Worldwide from 2012 to 2018 (in Million Head)," https://www.statista.com/statistics/263979/global-cattle-population-since-1990.
15. "Counting Chickens," The Economist, July 27, 2011.
16. S. G. Wiedemann, M.-J. Yan, B. K. Henry, and C. M. Murphy, "Resource Use and Greenhouse Gas Emissions from Three Wool Production Regions in Australia," Journal of Cleaner Production 122 (May 20, 2016): 121–132.
17. Kate Abnett, "Solving the Cashmere Crisis," Business of Fashion, November 26, 2015.
18. Jacob.

19. Haner et al.
20. Ministry of the Environment (Japan), "Climate Change in Mongolia: Outputs from GCM," 2015; NASA, "World of Change: Global Temperatures," https://earthobservatory.nasa.gov/WorldOfChange/DecadalTemp.
21. Schmitz.
22. Jacob.
23. Sarah J. Wachter, "Pastoralism Unraveling in Mongolia," *New York Times*, December 8, 2009. https://www.nytimes.com/2009/12/08/business/global/08iht-rbogcash.html
24. Qi Feng, Hua Ma, Xuemei Jiang, Xiu Wang, and Shixiong Cao, "What Has Caused Desertification in China?" *Scientific Reports* 5, article no. 15998 (2015).
25. Abnett.
26. Schmitz, NPR.
27. Siegle.
28. Osnos.

Fuel

1. https://www.epa.gov/ghgemissions/sources-greenhouse-gas-emissions
2. James D. Ward, Paul C. Sutton, Adrian D. Werner, et al., "Is Decoupling GDP Growth From Environmental Impact Possible?" *PLOS ONE*, October 14, 2016.
3. Lara P. Clark, Dylan B. Millet, Julian D. Marshall, "Changes in Transportation-Related Air Pollution Exposures by Race-Ethnicity and Socioeconomic Status: Outdoor Nitrogen Dioxide in the United States in 2000 and 2010," *Environmental Health Perspectives*, Vol. 125, No. 9, September 14, 2017.

The Other Problem with Coal

1. EPA, "Coal Ash Basics." In 2014, 130 million tons of coal ash were generated in the US.
2. Earthjustice, "Ash in Lungs: How Breathing Coal Ash Is Hazardous to Your Health," 2014.
3. Environmental Protection Agency, "Coal Ash Basics," https://www.epa.gov/coalash/coal-ash-basics.
4. Earthjustice, Environmental Integrity Project, Sierra Club, "Coal Ash: Seven Myths the Utility Industry Wants You to Believe and Seven Facts You Need to Know," 2011.
5. EPA.
6. EPA, Steam Electric Power Generating Effluent Guidelines, 2015; Earthjustice, "New EPA Data Show Coal Ash Problem Much Worse," June 27, 2012.
7. Southern Environmental Law Center, "Coal Ash," https://www.southernenvironment.org/cases-and-projects/coal-waste.
8. EPA, "EPA Response to Kingston TVA Coal Ash Spill," https://www.epa.gov/tn/epa-response-kingston-tva-coal-ash-spill.

9. EPA, "Duke Energy Coal Ash Spill in Eden, NC: History and Response Timeline," https://www.epa.gov/dukeenergy-coalash/history-and-response-timeline.

10. Jonathan M. Katz, "Duke Energy Is Charged in Huge Coal Ash Leak," *New York Times*, February 20, 2015.

11. EPA, "Case Summary: Duke Energy Agrees to $3 Million Cleanup for Coal Ash Release in the Dan River," https://www.epa.gov/enforcement/case-summary -duke-energy-agrees-3-million-cleanup-coal-ash-release-dan-river.

12. Department of Justice, "Duke Energy Subsidiaries Plead Guilty and Sentenced to Pay $102 Million for Clean Water Act Crimes," May 14, 2015, https://www .justice.gov/opa/pr/duke-energy-subsidiaries-plead-guilty-and-sentenced-pay -102-million-clean-water-act-crimes.

13. Mary Anne Hitt, "Coal Ash Was a Disaster in North Carolina Well before Hurricane Florence—and Now It's Even Worse," Sierra Club, October 1, 2018.

14. James Bruggers, "In Hurricane Florence's Path: Giant Toxic Coal Ash Piles," *Inside Climate News*, September 12, 2018.

15. Tatiana Schlossberg, "2 Tennessee Cases Bring Coal's Hidden Hazard to Light," *New York Times*, April 15, 2017.

16. Earthjustice, "New EPA Data Show Coal Ash Problem Much Worse," June 27, 2012.

17. Environmental Protection Agency, "Environmental Assessment for the Effluent Limitations Guidelines and Standards for the Steam Electric Power Generating Point Source Category," September 2015, pp. 3–47, https://www.epa.gov/sites /production/files/2015-10/documents/steam-electric-envir_10-20-15.pdf

18. Jennifer S. Harkness, Barry Sulkin, and Avner Vengosh, "Evidence for Coal Ash Ponds Leaking in the Southeastern United States," *Environmental Science and Technology* 50, no. 12 (2016): 6583–6592.

19. Environmental Protection Agency.

20. Physicians for Social Responsibility and Earthjustice, "Coal Ash: The Toxic Threat to Our Health and Environment," September 2010.

21. Schlossberg.

22. Thomas Cmar, interview with author, July 30, 2018.

23. US Energy Information Administration, "Frequently Asked Questions: What Is U.S. Electricity Generation by Energy Source?" https://www.eia.gov/tools/faqs/ faq.php?id=427.

24. US Energy Information Administration, "Coal Is the Most-Used Electricity Generation Source in 18 States; Natural Gas in 16," September 10, 2018, https://www .eia.gov/todayinenergy/detail.php?id=37034.

25. Steve Ahillen and Jake Lowary, "Judge Deals 'Huge Win' to Environmentalists in Federal Suit against TVA over Coal Ash in Gallatin," *Tennessean*, August 4, 2017.

26. Jonathan Mattise, "Appeals Court Reverses Tennessee Coal Ash Cleanup Order," Associated Press, September 24, 2018.

27. Shaila Dewan, "Clash in Alabama over Tennessee Coal Ash," *New York Times*, August 29, 2009.

28. Sabrina Shankman, "Alabama Town That Fought Coal Ash Landfill Wins Settlement," Inside Climate News, February 8, 2017.

29. Oliver Milman, "Environmental Racism Case: EPA Rejects Alabama Town's Claim over Toxic Landfill," *The Guardian*, March 6, 2018.

30. Jamie Satterfield, "Kingston Coal Ash Spill Worker: 'They Told Us We Would Be Fired If We Wore a Mask'—and He Was," *Knoxville News Sentinel*, October 19, 2018.

31. Jamie Satterfield, "Jury: Jacobs Engineering Endangered Kingston Disaster Clean-Up Workers," *Knoxville News Sentinel*, November 7, 2018.

32. EPA, Disposal of Coal Combustion Residuals from Electric Utilities, 2015.

33. EPA, "EPA Finalizes First Amendments to the Coal Ash Disposal Regulations Providing Flexibilities for States and $30M in Annual Cost Savings," July 18, 2018, https://www.epa.gov/newsreleases/epa-finalizes-first-amendments-coal-ash-disposal-regulations-providing-flexibilities.

34. Thomas Cmar.

The Wood for the Trees

1. Department for Business, Energy, and Industrial Strategy, "Implementing the End of Unabated Coal by 2025: Government Response to Unabated Coal Closure Consultation," January 2018.

2. European Union, "Renewable Energy Directive," https://ec.europa.eu/energy/en/topics/renewable-energy/renewable-energy-directive.

3. Emily Gosden, "Old King Coal's Reign at Drax 'Could End as Quickly as 2020,'" *Sunday Times*, September 14, 2017.

4. Katrin Bennhold, "For First Time since 1800s, Britain Goes a Day without Burning Coal for Electricity," *New York Times*, April 21, 2017.

5. Nathalie Thomas, "Drax Plans Huge Gas Plant and Battery Facility in North Yorkshire," *Financial Times*, September 13, 2017.

6. EPA, "EPA's Treatment of Biogenic Carbon Dioxide Emissions from Stationary Sources that Use Forest Biomass for Energy Production," April 18, 2018.

7. European Environment Agency, Scientific Committee, "Opinion of the EEA Scientific Committee on Greenhouse Gas Accounting in Relation to Bioenergy," September 15, 2011.

8. Timothy D. Searchinger, Steven P. Hamburg, Jerry Melillo, et al., "Fixing a Critical Climate Accounting Error," *Science* 326, September 23, 2009.

9. John Upton, "Pulp Fiction: The European Accounting Error That's Warming the Planet," *Climate Central*, October 20, 2015.

10. European Parliament, "Parliamentary Questions: Answer Given by Mr. Arias Cañete on Behalf of the Commission," http://www.europarl.europa.eu/doceo/document/E-8-2015-008770-ASW_EN.html.

11. Simon Evans, "Investigation: Does the UK's Biomass Burning Help Solve Climate Change?" Carbon Brief, May 11, 2015.

12. Upton.

13. S. R. Mitchell, M. E. Harmon, and K. E. B. O'Connell, "Carbon Debt and Carbon Sequestration Parity in Forest Bioenergy Production," *Global Change Biology Bioenergy* 4, no. 6 (2012): 818–827.

14. Upton.

15. Upton.

16. Upton.

17. Upton.

18. Justin Scheck and Ianthe Jeanne Dugan, "Europe's Green-Fuel Search Turns to America's Forests," *Wall Street Journal*, May 27, 2013. https://www.wsj.com/articles/SB10001424127887324082604578485491298208114

19. Oregon Fish and Wildlife Office, "Northern Spotted Owl," https://www.fws.gov/oregonfwo/articles.cfm?id=149489595.

20. Craig Hanson, Logan Yonavjak, Caitlin Clarke, et al., "Southern Forests for the Future," World Resources Institute, 2010.

21. M. C. Hansen, P. V. Potapov, R. Moore, et al., "High-Resolution Global Maps of 21st-Century Forest Cover Change," *Science* 342, no. 6160 (2013), http://science.sciencemag.org/content/342/6160/850.

22. Environmental Integrity Project, "Dirty Deception: How the Wood Biomass Industry Skirts the Clean Air Act," 2018.

23. US Energy Information Administration, "Biomass Explained: Wood and Wood Waste," https://www.eia.gov/energyexplained/index.php?page=biomass_wood.

24. Justin Gillis, "Let Forest Fires Burn? What the Black-Backed Woodpecker Knows," *New York Times*, August 6, 2017.

25. Matt Richtel and Fernanda Santos, "Wildfires, Once Confined to a Season, Burn Earlier and Longer," *New York Times*, April 12, 2016.

26. Tatiana Schlossberg, "Climate Change Blamed for Half of Increased Forest Fire Danger," *New York Times*, October 10, 2016.

27. Justin Gillis, "In New Jersey Pines, Trouble Arrives on Six Legs," *New York Times*, December 1, 2013.

28. EPA, "Sources of Greenhouse Gas Emissions," https://www.epa.gov/ghgemissions/sources-greenhouse-gas-emissions.

29. Clinton N. Jenkins, Kyle S. Van Houtan, Stuart L. Pimm, and Joseph O. Sexton, "US Protected Lands Mismatch Biodiversity Priorities," *Proceedings of the National Academy of Sciences* 112, no. 16 (April 2015): 5081–5086.

30. Solomon Hsiang, Robert Kopp, Amir Jina, et al., "Estimating Economic Damage from Climate Change in the United States," *Science* 356, no. 6345 (June 30, 2017): 1362–1365.

31. Craig Hanson, John Talberth, and Logan Yonavjak, "Forests for Water: Exploring

Payments for Watershed Services in the U.S. South," World Resources Institute, February 2011.

Staying Cool, Getting Hotter

1. Coral Davenport, "Nations, Fighting Powerful Refrigerant That Warms Planet, Reach Landmark Deal," *New York Times*, October 15, 2016.
2. Davenport; Project Drawdown, "Refrigerant Management," https://www.draw down.org/solutions/materials/refrigerant-management.
3. Paul Hawken, ed., *Drawdown: The Most Comprehensive Plan Ever Proposed to Reverse Global Warming* (New York: Penguin Books, 2017), 164.
4. US Department of State, "The Montreal Protocol on Substances That Deplete the Ozone Layer," https://www.state.gov/e/oes/eqt/chemicalpollution/83007.htm.
5. PBL Netherlands Environmental Assessment Agency, "Trends in Global CO_2 and Total Greenhouse Gas Emissions: 2017 Report," 2017.
6. Project Drawdown.
7. Davenport.
8. International Energy Agency, "The Future of Cooling: Opportunities for Energy-Efficient Air Conditioning," 2018.
9. US Energy Information Administration, "Air Conditioning and Other Appliances Increase Residential Electricity Use in the Summer," May 22, 2017; Department of Energy, "Air Conditioning," 2014, https://www.energy.gov/ energysaver/home-cooling-systems/air-conditioning.
10. Assessment and Standards Division Office of Transportation and Air Quality, US Environmental Protection Agency, "Draft Regulatory Impact Analysis: Proposed Rulemaking to Establish Light-Duty Vehicle Greenhouse Gas Emission Standards and Corporate Average Fuel Economy Standards," September 2009, pp. 5–19; author's calculation; Department of Energy, "Air conditioning."
11. Edward L. Glaeser and Kristina Tobio, "The Rise of the Sunbelt," Harvard University, John F. Kennedy School of Government, 2007.
12. "No Sweat," *The Economist*, January 5, 2013.
13. Stan Cox, *Losing Our Cool: Uncomfortable Truths about Our Air-Conditioned World (and Finding New Ways to Get through the Summer)* (New York: New Press, 2010).
14. 270 to Win, "Historical Presidential Elections," https://www.270towin.com/ historical-presidential-elections.
15. Alan Barecca, Karen Clay, Olivier Deschenes, et al., "Adapting to Climate Change: The Remarkable Decline in the US Temperature-Mortality Relationship over the Twentieth Century," *Journal of Political Economy* 124, no. 1 (January 5, 2016).
16. Jose Guillermo Cedeño Laurent, Augusta Williams, Youssef Oulhote, et al., "Reduced Cognitive Function during a Heat Wave among Residents of Non-Air-Conditioned Buildings: An Observational Study of Young Adults in the Summer of 2016," *PLOS Medicine* 15, no. 7 (July 10, 2018).

17. Stan Cox, "Your Air Conditioner Is Making the Heat Wave Worse," *Washington Post*, July 22, 2016.
18. James Hansen, Makiko Satoa, Reto Ruedy, et al., "Global Temperature in 2015," Earth Institute, Columbia University, January 19, 2016.
19. Kendra Pierre-Louis and Nadja Popovich, "Nights Are Warming Faster than Days. Here's Why That's Dangerous," *New York Times*, July 11, 2018.
20. International Energy Agency, p. 11.
21. F. Salamanca, M. Georgescu, A. Mahalov, et al., "Anthropogenic Heating of the Urban Environment due to Air Conditioning," *Journal of Geophysical Research Atmospheres* 119, no. 10 (May 28, 2014): 5949–5965.
22. International Energy Agency, "Air Conditioning Use Emerges as One of the Key Drivers of Global Electricity-Demand Growth," May 15, 2018, https://www.iea.org/newsroom/news/2018/may/air-conditioning-use-emerges-as-one-of-the-key-drivers-of-global-electricity-dema.html.
23. International Energy Agency, "Future of Cooling," p. 3.
24. International Energy Agency, "Future of Cooling," p. 63.
25. International Energy Agency, "Future of Cooling," p. 11.
26. International Energy Agency, "Future of Cooling."
27. International Energy Agency, "Future of Cooling," p. 38.
28. Kendra Pierre-Louis, "The World Wants Air-Conditioning. That Could Warm the World," *New York Times*, May 15, 2018.
29. International Energy Agency, "Future of Cooling," p. 12.
30. International Energy Agency, "Future of Cooling," p. 13.

The Great Big Cargo Route in the Sky

1. Damian Paletta, "In Rose Beds, Money Blooms," *Washington Post*, February 10, 2018.
2. Paletta.
3. Brandon Graver, "Yes, Your Mother Loves the Flowers, but Maybe Not the Cost of Flying Them In," International Council on Clean Transportation, May 9, 2018.
4. Brandon Graver, "Yes, Your Mother Loves the Flowers, but Maybe Not the Cost of Flying Them In."
5. Paletta.
6. International Air Transport Association, "The Value of Air Cargo—Air Cargo Makes it Happen."
7. Boeing, "Boeing Forecasts Air Cargo Traffic Will Double in 20 Years," October 16, 2018, https://boeing.mediaroom.com/2018-10-16-Boeing-Forecasts-Air-Cargo-Traffic-Will-Double-in-20-Years.
8. Project Drawdown, "Airplanes," https://www.drawdown.org/solutions/transport/airplanes.

9. IATA, "IATA Forecast Predicts 8.2 Billion Air Travelers in 2037," October 24, 2018.

10. Project Drawdown.

11. "Airlines Are to Cough Up for Cross-Border Pollution," *The Economist*, October 10, 2016.

12. Anastasia Kharina and Daniel Rutherford, "Fuel Efficiency Trends for New Commercial Jet Aircraft: 1960 to 2014," International Council on Clean Transportation, 2015, p. iii.

13. Alex Davies, "Planes Have to Get More Efficient. Here's How to Do It," *Wired*, June 11, 2015.

14. Center for Climate and Energy Solutions, "A New Flight Path for Reducing Emissions from Global Aviation," October 6, 2016.

15. Jane E. Penner, David H. Liter, David J. Grigg, et al., eds., "IPCC Special Report: Aviation and the Global Atmosphere," 1995.

16. Zach Wichter, "Too Hot to Fly? Climate Change May Take a Toll on Air Travel," *New York Times*, June 20, 2017.

17. E. Coffel and R. Horton, "Climate Change and the Impact of Extreme Temperatures on Aviation," *Weather, Climate, and Society* 7 (2014).

18. Henry Fountain, "Over 190 Countries Adopt Plan to Offset Air Travel Emissions," *New York Times*, October 6, 2016.

19. Luke N. Storer, Paul D. Williams, and Manoj M. Joshi, "Global Response of Clear-Air Turbulence to Climate Change," *Geophysical Research Letters* 44, no. 19 (October 16, 2017).

20. Daniel Rutherford, "ICAO, Why Can't You Be a Bit More Like Your Sister?" International Council on Clean Transportation, May 2, 2018.

21. Anne Petsonk, "ICAO's Market-Based Measure Could Cover 80% of Aviation Emissions Growth in Mandatory Phase," Environmental Defense Fund, October 6, 2016.

22. Rutherford.

23. Annie Petsonk, "To Understand Airplanes' Climate Pollution, a Picture Is Worth a Thousand Words," Environmental Defense Fund, February 12, 2016.

24. Davies.

25. Project Drawdown.

26. Tatiana Schlossberg, "Flying Is Bad for the Planet. You Can Help Make It Better." *New York Times*, July 27, 2017.

27. CO_2 emissions (metric tons per capita), World Bank, https://data.worldbank.org/indicator/EN.ATM.CO2E.PC?year_high_desc=true.

28. John Wihbey, "Fly or Drive? Parsing the Evolving Climate Math," Yale Climate Connections, September 2, 2015.

29. Yoon Jung, "Fuel Consumption and Emissions from Airport Taxi Operation,"

Green Aviation Summit, September 8–9, 2010, https://flight.nasa.gov/pdf/18_jung_green_aviation_summit.pdf.

30. Heinrich Bofinger and Jon Strand, "Calculating the Carbon Footprint from Different Classes of Air Travel," World Bank, Development Research Group, Environment and Energy Team, May 2013.

31. Michael Slezak, "Qantas Worst Airline Operating Across Pacific for CO_2 Emissions, Analysis Reveals," The Guardian, January 16, 2018, https://www.theguardian.com/business/2018/jan/17/qantas-worst-airline-operating-across-pacific-for-co2-emissions-analysis-reveals.

32. Brandon Graver and Daniel Rutherford, "Transpacific Airline Fuel Efficiency Ranking, 2016," International Council on Clean Transportation, 2018; Brandon Graver and Daniel Rutherford, "Transatlantic Airline Fuel Efficiency Ranking, 2016," International Council on Clean Transportation, 2018.

33. Malte Humpert, interview with author, October 16, 2018.

Shipping: The World in a Box

1. Joseph Stromberg, "The MSC Oscar Just Became the World's Biggest Container Ship," Vox, January 8, 2015.

2. Stromberg.

3. Edward Hume, Door to Door: The Magnificent, Maddening, Mysterious World of Transportation (New York: Harper Collins, 2016), Kindle edition, p. 180.

4. Zheng Wan, Shen Chen, and Daniel Sperling, "Pollution: Three Steps to a Green Shipping Industry," Nature News 530 (February 17, 2016): 275–277.

5. Wan, Chen, and Sperling.

6. Mike Schuler, "OOCL Hong Kong Breaks 21,000 TEU Mark, Becoming 'World's Largest Containership'," GCaptain, May 15, 2017, https://gcaptain.com/oocl-hong-kong-breaks-21000-teu-mark.

7. Hume, p. 30.

8. Hume, p. 181.

9. Wan, Chen, and Sperling.

10. Naya Olmer, Brian Comer, Biswajoy Roy, et al., "Greenhouse Gas Emissions from Global Shipping, 2013–2015," International Council on Clean Transportation, October 2017, p. 2.

11. Olmer et al., pp. iv, 2.

12. Hume, p. 180; Bryan Comer and Naya Olmer, "Heavy Fuel Oil Is Considered the Most Significant Threat to the Arctic. So Why Isn't It Banned Yet?" International Council on Clean Transportation, September 15, 2016, https://www.theicct.org/blogs/staff/heavy-fuel-oil-considered-most-significant-threat-to-arctic.

13. Hume, p. 180.

14. Mireya Navarro, "City Issues Rule to Ban Dirtiest Oils at Buildings," *New York Times*, April 21, 2011.

15. Olmer et al., p. viii.

16. Natural Resources Defense Council, "Prevention and Control of Shipping and Port Air Emissions in China," February 2014, pp. 12, 7, 11; Tami C. Bond et al., "Black Carbon in the Climate System: A Scientific Assessment," *Journal of Geophysical Research: Atmospheres* 118 (June 2013): 5380–5552.

17. Wan, Chen, and Sperling.

18. Natural Resources Defense Council, p. 12.

19. James J. Corbett, Paul S. Fischbeck, and Spyros N. Pandis, "Global Nitrogen and Sulphur Inventories for Ocean-Going Ships," *Journal of Geophysical Research* 104, no. 3 (February 1999): 3457–3470.

20. Wan, Chen, and Sperling.

21. Natural Resources Defense Council, p. 19.

22. Julia Pyper, "EPA Bans Sooty Ship Fuel off U.S. Coasts," *Climatewire, E&E News*, August 2, 2012.

23. Hume, p. 181.

24. International Energy Agency, "International Maritime Organization Agrees to First Long-Term Plan to Cut Emissions," April 13, 2018,

25. International Maritime Organization, "Sulphur 2020: Cutting Sulphur Oxide Emissions," http://www.imo.org/en/MediaCentre/HotTopics/Pages/Sulphur-2020.aspx.

26. Wan, Chen, and Sperling.

27. Bryan Comer and Natalie Olmer, "Black Carbon: Bringing the Heat to the Arctic," International Council on Clean Transportation, September 9, 2016.

28. Center for Climate and Energy Solutions, "Black Carbon Factsheet," 2020.

29. Center for Climate and Energy Solutions.

30. Center for Climate and Energy Solutions.

31. International Maritime Organization.

32. International Council on Clean Transportation, "Prevalence of Heavy Fuel Oil and Black Carbon in Arctic Shipping, 2015–2025," May 1, 2017.

33. Henry Fountain, "With More Ships in the Arctic, Fears of Disaster Rise," *New York Times*, July 23, 2017.

34. Clay Dillow, "Russia and China Vie to Beat the US in the Trillion-Dollar Race to Control the Arctic," CNBC.com, February 6, 2018.

35. Dillow.

36. "The Thawing Arctic Threatens an Environmental Catastrophe," *The Economist*, April 29, 2017.

37. "China Wants to Be a Polar Power," *The Economist*, April 14, 2018.

38. Comer and Olmer.

39. UN Environment Programme, "Integrated Assessment of Black Carbon and Tropospheric Ozone Summary for Decision Makers," 2011.

40. UN Environment Programme.

Cars, Trucks, and Justice

1. Brad Plumer, "Power Plants Are No Longer America's Biggest Climate Problem. Transportation Is," Vox, June 13, 2016.
2. Environmental Protection Agency, "Fast Facts on Transportation Greenhouse Gas Emissions," https://www.epa.gov/greenvehicles/fast-facts-transportation-greenhouse-gas-emissions.
3. Adie Tomer, Robert Puentes, and Joseph Kane, "Metro-to-Metro: Global and Domestic Goods Trade in Metropolitan America," Global Cities Initiative: A Joint Project of Brookings and JP Morgan Chase, 2013.
4. Brady Dennis, "White House Sets New Fuel-Efficiency Standards for Heavy-Duty Trucks, Vans, and Buses," Washington Post, August 16, 2016.
5. Nic Lutsey, "Eighteen Wheels and Ten Miles per Gallon," International Council on Clean Transportation, April 21, 2015.
6. Lisa Friedman, "New E.P.A. Chief Closes Dirty-Truck Loophole Left by Scott Pruitt," New York Times, July 27, 2018.
7. Friedman.
8. Friedman.
9. Eric Lipton, "'Super Polluting' Trucks Receive Loophole on Pruitt's Last Day," New York Times, July 6, 2018.
10. Friedman.
11. Eric Miller, "EPA's Andrew Wheeler Says New Glider Truck Rule Would Consider Impact on Manufacturers," Transport Topics, February 1, 2019. https://www.ttnews.com/articles/epas-andrew-wheeler-says-new-glider-truck-rule-would-consider-impact-manufacturers.
12. Environmental Defense Fund, "Green Freight Facts and Figures," http://business.edf.org/projects/green-freight-facts-figures.
13. Edward Hume, Door to Door: The Magnificent, Maddening, Mysterious World of Transportation (New York: Harper Collins, 2016) Kindle edition, p. 6.
14. Climate Central, "Transportation Is the Biggest Source of CO_2 in the U.S.," November 21, 2017.
15. Bill Vlasic, "U.S. Sets Higher Fuel Efficiency Standards," New York Times, August 28, 2012.
16. David Roberts, "Trump Is Freezing Obama's Fuel Economy Standards. Here's What That Could Do," Vox, August 2, 2018.
17. US Department of Energy, "Where the Energy Goes: Gasoline Vehicles," https://www.fueleconomy.gov/feg/atv.shtml.
18. Michael Kimmelman, "Paved, but Still Alive," New York Times, January 6, 2012.
19. US Department of Transportation, Federal Highway Administration.
20. Madeleine Rubenstein, "Emissions from the Cement Industry," Columbia University Earth Institute, May 9, 2012.
21. Environmental Protection Agency.

22. American Lung Association, "Nitrogen Dioxide," https://www.lung.org/our -initiatives/healthy-air/outdoor/air-pollution/nitrogen-dioxide.html.

23. Lara P. Clark, Dylan B. Millet, and Julian D. Marshall, "Changes in Transportation-Related Air Pollution Exposures by Race-Ethnicity and Socioeconomic Status: Outdoor Nitrogen Dioxide in the United States in 2000 and 2010," *Environmental Health Perspectives* 125, no. 9 (September 14, 2017).

24. Paul Mohai, Paula Lantz, Jeff Morenoff, et al., "Racial and Socioeconomic Disparities in Residential Proximity to Polluting Industrial Facilities: Evidence from the Americans' Changing Lives Study," *American Journal of Public Health* 99, S3 (2009): S649–S656.

25. Department of Environment, Food, and Rural Affairs, "Air Pollution in the UK 2015," September 2016.

26. Daniela Fecht, Paul Fischer, Léa Fortunato, et al., "Associations between Air Pollution and Socioeconomic Characteristics, Ethnicity, and Age Profile of Neighbourhoods in England and the Netherlands," *Environmental Pollution* 198 (March 2015): 201–210.

Hitching a Ride(share)

1. Uber, "Taking 1 Million Cars off the Road in New York City," July 10, 2015, https://www.uber.com/blog/new-york-city/taking-1-million-cars-off-the -road-in-new-york-city.

2. "Tatiana Schlossberg, George Moran," *New York Times*, September 10, 2017, https://www.nytimes.com/2017/09/10/fashion/weddings/tatiana-schlossberg -george-moran.html.

3. Uber.

4. California Public Utilities Commission, "CPUC Finds Uber Technologies Is Both a TNC and a Charter Party Carrier," 2018, http://docs.cpuc.ca.gov/Published Docs/Published/G000/M213/K609/213609321.PDF.

5. US Department of Commerce, Economics and Statistics Administration, Census Bureau, "We Are Gathered Here: Percentage of Each State's Population Living in Incorporated Places," 2015.

6. Regina R. Clewlow and Gouri Shankar Mishra, "Disruptive Transportation: The Adoption, Utilization, and Impacts of Ride-Hailing in the United States," University of California, Davis, Institute of Transportation Studies, October 2017.

7. Uber, "One in a Billion," December 31, 2015, https://www.uber.com/newsroom/ one-in-a-billion.

8. Uber, "5 Billion Trips," June 29, 2017, https://www.uber.com/newsroom/5billion-2.

9. Uber, "10 Billion," July 24, 2018, https://www.uber.com/newsroom/10-billion.

10. Schaller Consulting, "The New Automobility: Lyft, Uber and the Future of American Cities," July 25, 2018.

11. Schaller Consulting.

12. Clewlow and Mishra.
13. Clewlow and Mishra.
14. Clewlow and Mishra.
15. Schaller Consulting.
16. Caroline Rodier, "Hailing Services on Travel and Associated Greenhouse Gas Emissions," National Center for Sustainable Transportation, University of California, Davis, Institute of Transportation Studies, April 2018.
17. Schaller Consulting.
18. Laura Bliss, "Uber and Lyft Could Do a Lot More for the Planet," *Citylab*, April 30, 2018.
19. Schaller Consulting.
20. Clewlow and Mishra.
21. Winnie Hu, "Your Uber Car Creates Congestion. Should You Pay a Fee to Ride?" *New York Times*, December 26, 2017.
22. Lyft, "Paving the Way for Greener Cities," April 22, 2015, https://blog.lyft.com/posts/earthday.
23. Ingrid Lunden, "Uber Says That 20% of Its Rides Globally Are Now on Uber-Pool," TechCrunch, May 10, 2016.
24. Schaller Consulting.
25. Clewlow and Mishra.
26. Emma G. Fitzsimmons, "Subway Ridership Declines in New York. Is Uber to Blame?" *New York Times*, February 23, 2017.
27. Clewlow and Mishra.
28. Schaller Consulting.
29. INRIX, Global Traffic Scorecard, http://inrix.com/scorecard.
30. Prashant Kumar and Anju Goel, "Concentration Dynamics of Coarse and Fine Particulate Matter at and around Signalised Traffic Intersections," *Environmental Science: Processes and Impacts* 18 (2016): 1220.
31. Andrew J. Hawkins, "Uber Will Start Paying Some Drivers to Switch to Electric Cars," *The Verge*, June 19, 2018.
32. Uber, "Electrifying Our Network," June 19, 2018, https://www.uber.com/newsroom/electrifying-our-network.
33. Hawkins.
34. Uber, "Welcome, JUMP!" April 9, 2018, https://www.uber.com/newsroom/welcomejump; Lyft, "Introducing Lyft Bikes," July 2, 2018, https://blog.lyft.com/posts/lyft-to-acquire-us-bikeshare-leader.
35. Uber, "Uber Air," https://www.uber.com/us/en/elevate.
36. Lew Fulton, Jacob Mason, and Dominique Meroux, "Three Revolutions in Global Transportation," University of California, Davis, Sustainable Transportation Energy Pathways of the Institute of Transportation Studies, 2017.
37. Emma G. Fitzsimmons, "Uber Hit with Cap as New York City Takes Lead in Crackdown," *New York Times*, August 8, 2018.

38. Schaller Consulting; Laura Bliss, "How to Fix New York's 'Unsustainable' Traffic Woes," *CityLab*, December 21, 2017.
39. Charles Komanoff, "Leave It to Lyft? Trust Uber? (But Who's Gonna Watch the FHV Surcharges?)" *Streetsblog NYC*, October 18, 2018, https://nyc.streetsblog.org/2018/10/18/leave-it-to-lyft-trust-uber-but-whos-gonna-watch-the-fhv-surcharges/.
40. Winnie Hu, "Over $10 to Drive in Manhattan? What We Know about the Congestion Pricing Plan," *New York Times*, March 26, 2019.

Conclusion

1. Northeast Seafood Coalition, "History of New England's Groundfish Fishery and Management," https://northeastseafoodcoalition.org/fishery-101/history; Kevin Miller, "Maine Objects, but Regulators Vote to Keep Shrimp Fishery Closed for 2018," *Portland Press Herald*, November 29, 2017, https://www.pressherald.com/2017/11/29/federal-regulators-extend-moratorium-on-shrimp-fishery; Penelope Overton, "Maine Lobstermen Say Move to Avert Collapse of Herring Fishery Will Have Dire Consequences," *Portland Press Herald*, September 25, 2018, https://www.pressherald.com/2018/09/25/maine-lobstermen-say-move-to-avert-collapse-of-herring-fishery-will-have-dire-consequences.

Afterword

1. Anthony Leiserowitz, Edward Maibach, Seth Rosenthal, et al., "Climate Change in the American Mind: April 2020," Yale Program on Climate Change Communication, May 19, 2020.
2. Kenneth V. Rosenberg, Adriaan M. Dokter, Peter J. Blancher, et al., "Decline of the North American Avifauna," *Science* 366, no. 6461 (October 4, 2019).
3. Leanne Martin, Mathew P. White, Annie Hunt, et al., "Nature Contact, Nature Connectedness, and Associations with Health, Wellbeing, and Pro-Environmental Behaviours," *Journal of Environmental Psychology* 68 (April 2020).

About the Author

Tatiana Schlossberg is a journalist writing about climate change and the environment. She previously reported on those subjects for the Science and Climate sections of the *New York Times*, where she also worked on the Metro desk. Her work has also appeared in the *Boston Globe*, the *Atlantic*, *Bloomberg View*, *Yale Environment 360*, and elsewhere. She lives in New York.